THE MODELS OF

SKILL

ACQUISITION

AND

EXPERTISE

DEVELOPMENT

A QUICK REFERENCE OF SUMMARIES

Dr. Raman K. Attri

A publication by Speed To Proficiency Research: S2Pro©

ISBN: 978-981-11-8988-3 (e-book)
ISBN: 978-981-14-1122-9 (paperback)
ISBN: 978-981-14-1130-4 (hardcover)

First published: 2019
Lead author: Raman K. Attri
Published by Speed To Proficiency Research: S2Pro©
Published at Singapore
Printed in the United States of America

National Library Board, Singapore Cataloguing in Publication Data

Names: Attri, Raman K., 1973-
Title: The models of skill acquisition and expertise development : a quick reference of summaries / Raman K. Attri.
Description: Singapore : Speed To Proficiency Research, [2019]
Identifiers: OCN 1055689270 | ISBN: 978-981-14-1130-4 (hardcover)
| ISBN: 978-981-14-1122-9 (paperback)
| ISBN 978-981-11-8988-3 (e-book) | Includes bibliographic references.
Subjects: LCSH: Learning ability. | Learning, Psychology of. | Employees--Training of.
Classification: DDC 658.312404--dc23

Ｓ Speed To Proficiency
RESEARCH

Speed To Proficiency Research: S2Pro©
Singapore
https://www.speedtoproficiency.com
contact@speedtoproficiency.com

To my friends and teachers at National Institute of Technology Jalandhar – a place of first stepping stone to my journey toward expertise development !!

Summaries of 23 models of skill acquisition and expertise development for human performance and professional mastery

CONTENTS

CHAPTER 1

LEARNING, SKILL ACQUISITION AND EXPERTISE DEVELOPMENT

CHAPTER 2

CLASSIFYING THE MODELS OF SKILL, PROFICIENCY AND EXPERTISE DEVELOPMENT

CHAPTER 3

STAGE-BASED MODELS

CHAPTER 4

PRACTICE-, TIME-, OR TASK-BASED MODELS

CHAPTER 5

FACTOR-BASED MODELS

CHAPTER 6

EXPERT MODELING-BASED MODELS

CHAPTER 7

COGNITION-BASED MODELS

CHAPTER 8

PHASES OF SKILL ACQUISITION: INTEGRATING VARIOUS VIEWS

PREFACE

Broadly the book deals with the discipline of education. It cross-cuts in the disciplines of learning sciences, training, instructional design, cognitive sciences, cognitive psychology, expertise development, and performance improvement.

In the context of organizations, researchers and corporate leaders have started focusing on how to ensure robust human resource development and performance management of employees to meet their business goals. For the last half a century, researchers and scholars have tried to unlock the mysteries of human learning, identify the process of developing people to a higher level of competence, and improve overall performance at the jobs and other areas. The knowledge base in these subject areas is usually organized as models, frameworks, theories, or explanations of how people acquire new skills and become an expert in those skills. A vast amount of literature has been written about major models mentioned in most foundational research papers. This book summarizes several foundational models and theories from the literature review of four-decade of research on learning, skill acquisition, expertise development, and performance improvement of people.

HR specialists, learning designers, instructional designers, and performance consultants are trying to structure their training and learning programs for employee development around established models or best practices. This book clarifies the mechanisms of skill acquisition and expertise development, which can be used to design research-based curriculums in educational or corporate settings.

This book is forked out from the literature review during my doctorate research program, which eventually changed direction to a different topic. However, I find that this literature review still holds value for new researchers in the discipline of education, training, and learning. In general, the researchers are required to demonstrate a broader understanding of key models, theories, and philosophies in their disciplines in the literature review section of their research proposal or research paper. Arguably, this is also the most challenging step, upon which most students and researchers are not trained. A typical new researcher in the field of learning, training, cognitive sciences, education, or allied disciplines starts painfully reading thousands of pages of scattered research papers written about various skill acquisition and expertise development models. After extensive analysis, they start forming their viewpoint and converge to a specific theory, model, or aspect of it as the philosophical base for their research proposal or further exploration on a topic.

Ironically, most new researchers do not get a good handle on where to start their literature review and which direction to go for more in-depth exploration. That is where this book may act as a time saver for those new scholars or research students in training, education, and learning disciplines who are either new to these disciplines or who do not have any prior experience in conducting a literature review in these disciplines.

This book presents new researchers in learning, training, cognitive sciences, or education disciplines with a big picture starting point for their literature review journey. This book aims to offer a collection of summaries of existing models and theories to give a bird's eye view to new researchers based on which they can decide which direction to dig further. The book provides them a condensed but 'just enough' review of 23 significant models of skill acquisition and expertise development presented by leading researchers during the last half a century of classic and new research in a single volume. The review is complemented with over 200 authentic sources, which a researcher read for a detailed and deeper dive and set the direction for further exploration.

Last but not least, I must admit that this book is not a detailed reference and does not replace the in-depth review required during research phases. It is probably relevant to mention that the book should be considered a collection of summaries derived from a literature review of several famous skill acquisition and expertise development models. This book does not intend to offer an in-depth critique of the existing models and does not attempt to derive any new model or new theory.

Ideally, this book is meant for research students, researchers, and scholars (new and seasoned) who are researching the areas of human performance, education, expertise, skill development, instructional design, cognitive sciences, learning theory, training design, and similar disciplines. Academics or professors teaching courses or subjects in the above areas may find this book useful enough to recommend it to their students to plan for detailed literature reviews in such areas.

This book may also be helpful to professional instructional designers, learning specialists, trainers, and consultants interested in designing and delivering their programs or training sessions. The book may help them to understand the mechanisms of human performance improvement, the process of human skill acquisition, and theories of human expertise development.

Lastly, please contact me with your proposal if you are a budding researcher or a research student who would like to co-author an expanded edition of this book. It could be an edition offering summaries/reviews of additional models or extending already covered models with expanded critiques from several different disciplines.

Raman K. Attri (Dr)

April 2019

♣ ♣ ♣ ♣

The journey toward expertise is unceasing. Even those who have attained the knowledge, strategic abilities, and interests indicative of expertise cannot sit idly by as the domain shifts under their feet.

Alexander (2003, p. 12)

THE BOOK

The book offers condensed summaries of twenty-three major skill acquisition and expertise development models presented by leading researchers during the last half a century of classic and new research. This book presents new researchers in learning, training, cognitive sciences, or education disciplines with a big picture starting point for their literature review journey.

The book presents an easy-to-understand taxonomy of twenty-three models, giving new researchers a good bird's eye view of existing models and theories. They can decide which direction to dig further. The reviews in this book are complemented with over 200 authentic sources, which a researcher read for a detailed and deeper dive and set the direction for further exploration. This book would also act as an essential reference for training & learning professionals and instructional designers to design research-based training curriculum to develop the skills of their staff.

Chapter 1 of the book elaborates on how the processes of learning, skill acquisition, and expertise development are interwoven.

Chapter 2 presents a classification system to categorize various models reviewed in the literature under five groups.

Chapter 3 describes twelve models of skill and expertise acquisition which are represented in the form of stages used frequently in learning, training, and performance literature. The chapter also briefly discusses each model's implications toward developing the skills and expertise of a less proficient individual to a higher level of proficiency.

Chapter 4 reviews practice-, time- or task-based models, which are theories or models suggesting that acquisition of knowledge & skills, development of expertise, and performance improvement is a function of nature of the practice, amount of time spent on the task and task type.

Chapter 5 presents the factor-based models, suggesting the interplay of several factors that influence the acquisition of knowledge & skills, development of expertise, and performance improvement.

Chapter 6 embarks on describing expert modeling-based models, suggesting modeling an expert through elicitation or guidance for the acquisition of knowledge & skills, development of expertise, and performance improvement.

Chapter 7 covers some newer movements toward cognition-based models, which are theories or models focusing on mechanisms of cognitive learning for the acquisition of knowledge & skills, development of expertise, and performance improvement.

Chapter 8 concludes the book by integrating views from various thought leaders to explain a famous staged skill acquisition model.

♣ ♣ ♣ ♣

THE AUTHOR

Dr Raman K Attri is a corporate business researcher, learning strategist, and professional speaker with a strong zeal to enable people to unravel human learning and performance. He specializes in providing competitive and strategic value to organizations by accelerating the time-to-proficiency of employees through well-researched models. He holds a doctorate in business from Southern Cross University, Australia. His international professional career spanned over 25 years across various disciplines, such as scientific research, systems engineering, management consulting, training operations, professional teaching, and learning design. A strong proponent of learning as the core of human success, he provides advisory on accelerated learning techniques which earned him over 100 educational credentials, including two doctorate degrees, three masters' degrees, and tens of international certifications. Despite physical disability since childhood, he leveraged it to learn, research, and test a range of "how-to methods" to accelerate the rate of personal learning and professional performance at the workplace. He has published his methods in scholarly journals, blogs, books, and conferences. He also runs a "Get There Faster" movement to enable people to speed up their mastery. He can be reached at http://ramankattri.com.

♣ ♣ ♣ ♣

ABBREVIATIONS

ACT-R	Adaptive control of thought—rational
CDM	Critical decision method
CFT	Cognitive flexibility theory
CTA	Cognitive task analysis
CTT	Cognitive transformation theory
EPRI	Electric Power Research Institute
GOMS	Goals, operators, methods, selections
HTA	Hierarchical task analysis
ITAM	Integrated task analysis model
MDL	Model of domain learning
RPD	Recognition-primed decision
RWL	Researching in work and learning
WDA	Work domain analysis
ZPD	Zone of proximal development
TTP	Time to proficiency

CHAPTER 1

LEARNING, SKILL ACQUISITION AND EXPERTISE DEVELOPMENT

The processes of learning, skill acquisition, and expertise development are intricately interwoven when the discussion is about an employee's job performance. However, these terms do not always get their due distinction. On a broader scale, learning is assumed to be the most fundamental process in an individual's overall development. At the same time, it is also believed that learning may not always translate into performance at a job. While skill acquisition processes underscore learning as the most fundamental mechanism in an individual's overall development, expertise development is viewed as a journey in which learning and skill acquisition play an undeniable role in delivering the required performance.

1.1   LEARNING AND PERFORMANCE

In general, task performance and job performance are considered to be direct functions of the abilities or skills acquired by an individual. Hunter (1986), in one of his studies, reported from over 3000 cases that abilities impact job knowledge and work samples (performance), which, in turn, affect the supervisory ratings of performance. He noted that abilities did not have a direct effect on how employee supervisors rated them. Rather, the abilities impacted how job knowledge and work samples were accomplished. This theory supports that knowledge and skills are mediating variables in an individual's performance. McDaniel, Schmidt & Hunter (1988; 1986) suggested that knowledge, skills, and abilities are the primary determinants of job performance. Kanfer & Kantrowitz (2002, p. 32), based on an analysis of several studies, observed that 'individual differences in general cognitive ability may account for more variance in performance when performance is defined in terms of skill acquisition or job proficiency.'

Subsequently, several classical studies established that the knowledge, skills, and abilities acquired by an individual as a result of job experience or training are the primary determinants of his/her performance in any job (McDaniel, Schmidt & Hunter 1988; Qui'nones, Ford & Teachout 1995). Campbell (1990) explained that job performance could be distinguished from one person to another with the help of three direct determinants—declarative knowledge (knowledge of facts, principles, and procedures), procedural knowledge and skills (knowing what to do and actually doing the task), and motivation (choice to exert effort, how much and how long). The level of training, education, experience, range of skills, and amount of practice determine the level of declarative and procedural knowledge or skills possessed by an individual. This study supports the link between knowledge, skills, abilities, and experience toward individual performance.

Motowidlo, Borman & Schmit (1997) presented a model of task performance and contextual performance in which they theorized that the intervening variables to performance are knowledge, skills, and work habits, which are basically acquired through experience. They postulated that task knowledge (procedures, judgment, heuristics, rules, and decisions), task skills (using technical information, solving problems, and making judgments etc.), and task work habits (pattern of behaviors, tendencies, choices individuals make in a situation, motivational aspects, persistence, and planning, etc.) affect the task performance. Similarly, contextual knowledge (knowledge about the effective actions in situations that call for volunteering, helping, supporting, persisting, and defending etc.), contextual skills (ability to carry out actions deemed effective in a situation), and contextual work habits (ways of handling conflicts, tendencies, and interpersonal styles etc.) determine the contextual performance.

Thus, each of these kinds of performance (i.e., task performance, job performance, and contextual performance) is a direct function of the knowledge, skills, and abilities of an individual/employee, though it may not be entirely attributed to them. Several studies have proposed that learning has a central role in the performance concept and is an underlying mechanism in the acquisition of knowledge and skills (Campbell 1990; Hesketh & Neal 1999; London & Mone 1999). Learning is viewed as a long-term behavioral change, which positively impacts performance. Some studies show that individual performance improves with learning in terms of the time spent on the job (Avolio, Waldman & McDaniel 1990; McDaniel, Schmidt & Hunter 1988; Qui'nones, Ford & Teachout 1995). Learning a task or skill leads to ultimate performance, be it behavioral, task, outcome, or job performance. Thus, learning is considered a significant dimension of performance. Though the role of learning in improving performance is well understood, the nature of the relationship between the two has been debated extensively. For example, performance and learning share a direct relationship during training and an inverse relationship after training. During any training intervention, learning is contended to be more important than in-training performance, while after the training, at-the-job performance is contended to be more important than learning. In a study, Bjork (2009, p. 313) expressed that performance during a training event may not be the right indicator of job performance:

> Performance during training is often an unreliable guide to whether the desired learning has actually happened. Considerable learning can happen across periods when performance is not improving and, conversely, little or no learning can happen across periods when performance is improving markedly.

From this perspective, learning becomes more important than performance during training. Soderstrom & Bjork (2015, p. 193) termed learning as 'the relatively permanent changes in behavior or knowledge that support long-term retention and transfer' and performance during training as 'the temporary fluctuations in behavior or knowledge that are observed and measured during training or instruction or immediately thereafter.' Furthermore, the performance during a training intervention may not be the desired performance required at the job. However, what matters is the job performance of an individual after training. Bjork (2009, p. 319) highlights the challenge as: 'The problem for a training organisation is to maximise performance when it matters, that is, *after* training [sic] and, [sic] specially, when individuals are deployed.' The corollary to this assertion is that accelerating performance within training may have only short-term effects, which may be defined as 'expediting acquisition performance today does not necessarily translate into the type of learning that will be evident tomorrow' (Soderstrom & Bjork 2015, p. 193). However, the performance on the job (i.e., outcomes and behaviors) matters more than learning, as suggested by Sonnentag & Frese (2002, p. 6):

> One might argue that what ultimately counts for an organisation is the individuals' performance and not their learning—although learning might help to perform well. This line of reasoning stresses that learning is a highly relevant predictor of performance but is not performance itself.

A review of the literature on performance metrics indicates that learning is one of the key determinants of performance. A noteworthy observation is that knowledge and skill acquisition appear to be inseparable parts of job performance. In their study on pilots tackling direct problems in their job, Dreyfus & Dreyfus (2005) noted that an individual/employee (in this case, a pilot) passed through five stages during the course of acquiring experience: novice, advanced beginner, competent, proficient, and expert. The representation was based on the assumption that performance or skill proficiency was a continuum in which novice was at one end and expert at the other. The overall goal of knowledge and skill acquisition is to develop a higher level of competence, proficiency, or expertise of an individual, which then translates into performance through their behaviors/actions or results.

1.2   EXPERTISE

Expertise has been studied from several different perspectives.

According to one of the perspectives, expertise can be defined from three dimensions—development of expertise, knowledge structures possessed by experts, and reasoning processes used by experts (Hoffman 1998). Novak (2011) expressed that expertise can be defined through four lenses: attributes, cognition, stages, and community. The "attributes" lens defines expertise based on several years of study in different fields, which reveals some of the ways in which experts operate. The "cognition" lens defines expertise based on cognitive studies of experts in fields such as chess, music, and sports, etc., which reveal how experts think and organize knowledge. The "stages" lens views expertise as a sort of progression of knowledge and skills. The "community" lens views expertise as a quality that emerges from the interactions individuals have with fellow individuals and the environment.

One stream of literature is almost entirely devoted to novice–expert differences. The pioneering research by De Groot (1965; 1966) and Chase & Simon (1973) on the differences in the performance of novices and experts in the game of chess has generated plenty of research studies (for instance, studies by Chi, Glaser & Farr 1988). Several studies reported some characteristics in which experts were different from a novice and have attempted to explain the nature of expertise. For instance, some researchers believed that experts within their respective domains are skilled, competent, and think in qualitatively different ways than novices (Anderson 2000; Chi, Glaser & Farr 1988). Klein (1998) describes that expert performance comes by virtue of an expert's ability to integrate information from a large array of accumulated experiences to assess a situation; select a course of action through recognition; and then assess the course of action through mental simulation. This is termed as an intuitive capability, which only experts are deemed to have. Thus, exclusivity is one of the features of expertise that sets it apart from other constructs in skill acquisition. Expertise typically has been viewed as the abilities possessed by only a few people and usually are not common enough to be possessed by all (Dror 2011). The abilities may comprise a range of skills, knowledge, and performance characteristics and may vary among domains. Dror (2011, p. 179) summarized the capabilities of experts that help them achieve high-performance levels as:

> Experts need to have well-organized knowledge, use sophisticated and specific mental representations and cognitive processing, apply automatic sequences quickly and efficiently, be able to deal with large amounts of information, make sense of signals and patterns even when they are obscured by noise, deal with low quality and quantity of data, or with ambiguous information and many other challenging task demands and situations that otherwise paralyse the performance of novices.

A continuum view of the novice-to-expert transition has usually been used to explain the characteristics and actions of experts. Based on the continuum, Dreyfus & Dreyfus (2005) contend that an expert operates and behaves differently from a novice, advanced beginner, competent, or proficient performer. An expert exhibits experience-based deep understanding. According to Dreyfus and Dreyfus, 'An immense library of distinguishable situations is built up on the basis of experience' (Dreyfus & Dreyfus 1986, p. 32). Thus, experts treat knowledge in context, and possess the ability to recognize the relevance. Experts do not apply rules or use maxims or guidelines. Dreyfus and Dreyfus suggest that, individuals, at an expert level, rely on intuition and use an analytical approach only in new situations or while encountering unrecognized problems not experienced earlier. They showed that, at the expert stage, actions are rather effortless and out of intuition and tacit knowledge. Their problem-solving is based on an intuitive grasp of situations, the ability to recognize the features in a given situation, and a conceptual understanding of the underlying principles that govern those situations or occurrences. Expert performers possess the ability to see what needs to be achieved and how to achieve it. An expert 'focuses in on the accurate region of the problem without wasteful consideration of a larger range of unfruitful possibilities' (Benner 1984, p. 34). Similarly, experts possess the capacity to make subtle discriminations that proficient performers can not, which enables them to adapt their problem-solving approach based on the situation. Experts, based on their prior experience, can even come up with a solution for situations they have never experienced before (DiBello & Missildine 2011). They can see alternative approaches in a given situation. At this stage, their skills become so automatic that even they are not always conscious of the fact. Their performance is fluid. Therefore, it is believed that experts could move effortlessly between intuitive and analytical approaches and possess the ability to see the overall picture.

In almost all the expertise and proficiency development models, proficiency progression has been considered to be unidimensional, especially in staged models. Such approaches view proficiency as a trait possessed by an individual. However, researchers appear to recognize that experts develop 'specialist tacit knowledge' by becoming socially embedded in a group of experts (Collins 2011, p. 255). Collins (2011) and Collins *et al.* (2006) proposed the construct of interactional expertise. They proposed two additional dimensions to expertise: (1) the dimension that deals with the degree of exposure to tacit knowledge and (2) the dimension that deals with *esotericity* (specialized knowledge). They argue that it creates a three-dimensional *expertise space* in which expertise can be explored in a number of ways. The postulation by Collins (2011, p. 255) is based on the fundamental premise that:

> The acquisition of nearly every expertise, if not all of them, depends on the acquisition of the tacit knowledge pertaining to the expert domain in question. Tacit knowledge can be acquired only by immersion in the society of those who already possess it.

1.3   EXPERTISE DEVELOPMENT

In literature, skill acquisition and expertise development have been explained from different viewpoints. Most of the explanation of knowledge and skill acquisition has been developed from well-regarded classical and seminal work conducted during the 1960s, 1970s, and 1980s, such as Fitts & Posner (1967), Anderson (1981, 1982), Ackerman (1988), and Shiffrin & Schneider (1977). Some researchers in more recent times, such as Langan-Fox *et al.* (2002), have further expanded these theories. For several decades, one of the most common trends was describing skill acquisition in terms of stages in which performance or the proficiency of skills improved as individuals gained more practice and experience. Dreyfus & Dreyfus (2005) assert that an individual, during his/her course of acquiring experience, passes through five stages, namely, novice, advanced beginner, competent, proficient, and expert. The representation assumed that performance or skill proficiency is a continuum where the novice is at one end of the continuum while the expert is at the other.

Expertise is best understood by understanding what an expert does. Hoffman (1998, p. 85) defines an expert as:

> The distinguished or brilliant journeyman, highly regarded by peers, whose judgments are uncommonly accurate and reliable, whose performance shows consummate skill and economy of effort, and who can deal effectively with certain types of rare or "tough" cases. Also, an expert is one who has special skills or knowledge derived from extensive experience with subdomains.

From that perspective, expertise can be defined as 'the possession of a large body of knowledge and procedural skills' (Chi, Glaser & Rees 1982, p. 8). In their seminal collection, Glaser & Chi (1988) contested that there is a strong interplay between knowledge structures, processing capability, and problem-solving in the development of the desired expertise. The argument that expertise is being acquired, and, hence, is an outcome or goal of skill acquisition is posited by Chi (2006, p. 23) as 'presumably the more skilled person became expert-like from having acquired knowledge about a domain, that is, from learning and studying' and 'from deliberate practice.'

More number of research studies appear to support the view that developing a higher level of performance state, namely, proficiency and expertise, can be viewed as a process. Supporting this proposition, Lajoie (2003, p. 22) maintains that 'studies of expertise inform us that becoming an expert is a transitional process.' This development process is quite complex. An individual passes through several phases. 'The developmental process in expertise must be viewed as the complex process comprised of [sic] stagnation, retrogression, and growth. Rise in expertise should be understood in terms of the pursuing process [sic]' (Moon, Kim & You 2013, p. 228). Researchers maintain that expertise is not the ultimate goal; rather, it is a 'nonlinear process state' which means it is developed along the way in the form of stages (Moon, Kim & You 2013, p. 228). Bransford *et al.* (2004) also maintain that expertise is not a finished product; rather, it is an ongoing process. It appears that expertise is not a definite stage but grows out of interactions with the environment.

Further, possessing a certain amount of knowledge, skills, or behavior alone does not mean expertise. Instead, expertise is a continuous process of upgrading one's knowledge, relearning, changing one's mental representation as s/he encounters different situations, reflecting upon it, and using pattern recognition and correcting one's way of thinking.

Ge & Hardré (2010, p. 24) cited that 'Anderson (1985) and Dreyfus and Dreyfus (1986) redefined expertise by its processes and applications, such as the way in which experts solve problems, rather than simply by the amount of knowledge that experts possess.' In terms of the continuous process toward adjustments, Sternberg (1999, p. 359) contends that developing expertise involves 'the ongoing process of the acquisition and consolidation of a set of skills needed for a high level of mastery in one or more domains of life performance.' In the dynamic world, abilities are not always static. The nature of problems also is not always the same. Therefore, there is a need for the continuous acquisition and maintenance of expertise in the skills. Alexander (2003a, p. 12) makes a point that 'the journey toward expertise is unceasing. Even those who have attained the knowledge, strategic abilities, and interests indicative of expertise cannot sit idly by as the domain shifts under their feet.' Thus, it is reasonable to infer that expertise is a constant evaluation of one's zest toward mastery.

Several leading researchers have generated different theories and models to explain the process of expertise development and skill acquisition. Some research studies have focused on knowledge strategies and interactions of various factors that allow attention to transitions and trajectories to increase expertise. This book describes the major models/theories that explain skill acquisition and expertise development.

Chapter 2 of this book presents a taxonomy to classify a range of models and theories worth reviewing. The models are grouped thematically. The subsequent chapters of this book then dive into each thematic group in detail, summarizing each selected model under a given theme.

♣ ♣ ♣ ♣

CHAPTER 2

CLASSIFYING THE MODELS OF SKILL, PROFICIENCY AND EXPERTISE DEVELOPMENT

One challenge faced by most of the training and learning designers is developing curriculum strategies specifically directed at advancing learners toward a higher level of expertise in the skills acquired by them. The journey of a learner, from being a novice to an expert, is a fascinating topic. However, training professionals tend to follow or create unconfirmed theories on the topic. There are some theories with empirical evidence which attempt to explain the novice-to-expert transition and the way learners move from the novice stage to the expert stage. While doing this literature review, a comment from Fadde & Klein (2010, p. 5) guided me: 'As we search for methods of accelerating expertise, it is natural to look to theories of expertise and expert performance that have been developed through academic research and have recently entered public and business awareness as well.' In this book, I summarize the most popular and relevant models for developing a novice into an expert with a high level of proficiency and expertise.

The literature review suggested several theories and models for skill acquisition and expertise development. I propose to group these models under five thematic categories, as shown in the table. This grouping is based on the key foundational parameters of each model in terms of how they suggest skill acquisition or performance development. At a high level, this categorization can help researchers systematically organize the literature review.

Stage-based models: These include the theories and models that suggest the acquisition of knowledge and skills, development of expertise, and performance improvement occurring in the form of stages.

Practice-, time- or task-based models: These include the theories or models that suggest the progression of acquisition of knowledge and skills, development of expertise, and performance improvement occurring as a function of practice, time, and task.

Factor-based models: These include the theories or models that suggest the interplay of various factors in influencing the acquisition of knowledge and skills, development of expertise, and performance improvement.

Modeling-based models: These include the theories or models that suggest modeling an expert through elicitation or guidance as the key mechanism for acquiring knowledge and skills, development of expertise, and performance improvement.

Cognition-based models: These include the theories or models that focus on the mechanisms of cognitive learning for the acquisition of knowledge and skills, development of expertise, and performance improvement.

Table 1 presents the taxonomy of the major skill acquisition/expertise development models, which are described in detail in subsequent chapters. The models provide insights, guidance, and implications for training professionals.

Table 1: Classification of the major skill acquisition/expertise development models

Basis	Theory or Model
Stage-based models	Conscious competence theory
	Model of skill acquisition by Fitts & Posner (1967)
	ACT-R theory of cognition by Anderson (1982)
	Theory of ability determinants of skill acquisition by Ackerman (1988)
	Cognitive skill acquisition theory by VanLehn (1996)
	Theory of skill retention by Kim *et al.* (2013)
	Zone of proximal development by Vygtosky (1978)
	Staged skill acquisition model by Dreyfus & Dreyfus (1986)
	Model of domain learning by Alexander (1997)
	Proficiency scale by Hoffman (1998)
	Levels of human competence by Jacobs (1997)
	Novice-to-expert transition by Rosenberg (2013)
Practice-, time-, or task-based models	Mastery learning model by Bloom (1968)
	4-Component model of complex learning by van Merriënboer (1992)

	Deliberate Practice Model of Expert Performance by Ericsson *et al.* (1993)
	Recognition-Primed Decision (RPD) Model by Klein (1993)
Factor-based models	Model of Intelligence as Expertise Development by Sternberg (1999)
	Skilled Performance Theory by Langan-Fox *et al.* (2002)
Expert modeling-based models	Cognitive Apprenticeship Model by Collins (1989)
	Cognitive Task Analysis (CTA) Approach to Model Expertise
	The trajectories of Expertise by Lajoie (2003)
	Zone of proximal development by Vygtosky (1978)
Cognition-based models	Cognitive flexibility theory by Spiro *et al.* (1990)
	Cognitive Transformation Theory by Klein & Baxter (2009)
	ACT-R Theory of Cognition by Anderson (1982)

♣ ♣ ♣ ♣

CHAPTER 3

STAGE-BASED MODELS

A review of the literature suggests that the novice-to-expert transition can be viewed as a staged progression. The discussion is always an interesting one, but ironically, it never reaches a consensus in regard to the definitions and even names of different stages through which a novice develops into an expert. Though there are arguments against the existence of clear-cut stages, few studies have confirmed the occurrence of a level-like shift in some of the qualitative traits as a novice moves to become an expert. In this chapter, I will use the words 'stage' or 'phase' interchangeably.

Classical studies have defined staged progressions toward the automaticity of skills (Ackerman 1988; Anderson 1982; Fitts & Posner 1967). Subsequently, several researchers have developed more refined details of the stages toward proficiency. Among them, Shuell (1986, p. 364) noticed that 'as an individual acquires knowledge, his or her knowledge structure gradually evolves in qualitative as well as quantitative ways.' Classical studies indicated a qualitative shift in traits as novice learners moved toward higher proficiency (Benner

1984; Dreyfus & Dreyfus 1986). An example of such a shift is evident from the fact that master coaches can always anticipate the mistakes beforehand that learners are likely to make, depending on their skill levels. Similarly, the qualitative shifts were supported by Hoffman (1998, p. 84) as: 'The distinction between "novice" versus "expert" implies that development can involve both qualitative shifts and stabilizations in knowledge and performance.' These observations imply that the novice-to-expert transition can be viewed as being made up of several stages.

Theorists have used several developmental parameters for qualifying level-like shifts; for example, the transformation of skills as second nature, that is, automaticity (Ackerman 1988; Anderson 1982; Fitts & Posner 1967) as well as implicit knowledge (Alexander 2003b; Spiro *et al.* 2003). Some studies have established that it is possible to qualitatively define proficiency levels in terms of skills and knowledge exhibited by professionals on-the-job. The most significant studies in this regard were carried out by Dreyfus and Dreyfus (Dreyfus & Dreyfus 1986, 2004, 2005), popularly known as the *Dreyfus & Dreyfus model.* They contended that an individual passes through five stages of proficiency: novice, advanced beginner, competent, proficient, and expert. As one progresses through the stages, s/he becomes more intuitive toward solving problems. Benner (2004) attempted to define the levels of proficiency of the Dreyfus & Dreyfus model in terms of certain attributes of expected performance at each level in the context of the nursing and medical professions. Based on that, she demonstrated that progression could be reasonably explained using stages specified by the Dreyfus and Dreyfus model. Peña (2010) and Khan & Ramachandran (2012) applied the Dreyfus & Dreyfus model in the context of clinical practice and healthcare jobs, and asserted that there is a level-like progression that can be demarcated qualitatively in terms of job attributes.

Some researchers asserted that task performance could be an indicator of different stages (Chi 2006; Merrill 2006; Schreiber *et al.* 2009). According to this premise, a novice completes simple versions of his/her tasks during training, and moves to more complex tasks as his/her skill levels increase. Progressively, s/he becomes skillful at relatively more complex tasks and starts addressing several cues at the same time. Literature indicates that the measurement of task performance must reflect this gradual acquisition of skill. Merrill (2006, p. 269) stated that proficiency measurement requires one to 'detect increments in performance demonstrating gradually increased skill in completing a whole complex task or solving a problem.' Based on the above arguments, it was inferred that the novice-to-expert transition can be viewed as a staged progression.

Another group of researchers took a measurement approach to define proficiency levels in terms of the measurable attributes of jobs. For example, Chi (2006) proposed that proficiency levels could be roughly measured using inputs, such as academic qualification, years of experience in performing the task, peer feedback, and profession-related tests. Schreiber *et al.* (2009) developed metrics for the measurement of pilot proficiency using simulators and postulated that it is possible to develop an objective measurement of performance in complex jobs that require a range of judgment and meta-skills. Recently, Kim (2012, 2015) proposed a different approach, which involved measurement in terms of knowledge structures across different levels of the learning progress. Kim defined a set of measures corresponding to the levels of the features of the knowledge structure. The relationship between the measures and the features of knowledge structures was determined based on theoretical assumptions as well as empirical evidence. A similar approach was suggested by Dörfler, Baracskai & Velencei (2009), who used knowledge as the demarcation for levels to explain the stages or levels of expertise.

The concept of stages conveys the idea of progression being a process. Both foundational and more recent research studies support the idea that developing a higher level of performance to proficiency and expertise can be viewed as a process. While the basic idea of such a transition is that experts are relative to the novice, the goal of the approach is 'to understand how we can enable less skilled or experienced persons to become more skilled ...' (Chi 2006, p. 23).

This chapter describes the following 12 major models of skill acquisition and expertise development used frequently in learning, training, and performance literature. The chapter discusses the role of each model in developing the skills and expertise of less proficient employees, enabling them to attain a higher level of proficiency.

- Conscious competence theory by various

- Model of skill acquisition by Fitts & Posner (1967)

- Adaptive control of thought—rational (ACT-R) theory of cognition by Anderson (1982)

- Theory of ability determinants of skill acquisition by Ackerman (1988)

- Cognitive skill acquisition theory by VanLehn (1996)

- Theory of skill retention by Kim *et al.* (2013)

- Zone of proximal development by Vygotsky (1978)

- Staged skill acquisition model by Dreyfus & Dreyfus (1986)

- Model of domain learning by Alexandar (1997)

- Proficiency scale by Hoffman (1998)

- Levels of human competence by Jacobs (1997)

- Novice-to-expert transition by Rosenberg (2012)

3.1   CONSCIOUS COMPETENCE THEORY

When I was a kid, I got a new bicycle. Hardly able to wait even for a day, I attempted riding it, not worrying whether or not I knew how to ride. Not knowing how to peddle and when to steer, I fell down a couple of times. Despite my enthusiasm, these were enough to convince me that I did not know to ride a bicycle properly. As I took help and coaching from others, I started perfecting my riding and maneuvering the cycle in different situations. The more cycling practice I did, the more I was confident. Eventually, cycling was my second nature to the level that I was completely unconscious about the efforts I was exerting but highly competent while riding a bicycle. This is the essence of conscious competence (sometimes also called unconscious competence) theory.

The earliest known work in explaining the mechanism of proficiency development is the 'conscious competence theory.' Though the academic grounds of this theory are unknown, an article in *The Personnel Journal* (1974) attributes the origin of the theory to W. Lewis Robinson, Vice President of Industrial Training, International Correspondence Schools, USA (Robinson 1974)[1]. In this article, Robinson explains that an individual progresses through four specific stages (unconscious incompetence, conscious incompetence, conscious competence, and unconscious competence) of learning a new skill (behavior, ability, or technique) in common-sense terms. The original nomenclature used the term 'competency,' which was later replaced with the broader term 'competence'; the theory is called the 'conscious competence theory.'

---

[1] Interested readers may refer to a comprehensive account of the evolution of this model written by Alan Chapman at https://www.businessballs.com.

- **Unconscious incompetence:** A naïve is assumed to be at the 'unconscious incompetence' stage. At this stage, s/he is not aware of what skills s/he lacks, and, in several instances, does not even know the relevance or usefulness of the skills lacking.

- **Conscious incompetence:** The individual then moves to the 'conscious incompetence' stage, where s/he starts becoming aware of the existence and relevance of the required skills in a job and may start recognizing how much/what s/he does not know. Such recognition usually happens with a discovery process. Following this, s/he starts working toward learning those skill(s) that s/he lacked and does the necessary practice.

- **Conscious competence:** Eventually, the individual reaches the stage of 'conscious competence,' wherein s/he starts performing the learned skill reliably and without assistance. However, the individual still requires attention, and has not yet become automatic at it.

- **Unconscious competence:** However, with consistent practice and repeated use, the individual moves to the next stage of 'unconscious competence' where his ability to perform the skill becomes automatic (i.e., the skill enters the subconscious mind and becomes his/her second nature).

The 'conscious competence model' has several representations, the most popular one being the quadrant matrix representation (shown in Figure 1). The representation emphasizes that the progression is from quadrants 1 through 2 and 3 to 4. It is not possible to jump stages. For some advanced skills, an individual may actually regress back to the previous stage(s) for want of consistent practice on the newly learned skills (Chapman n.d.). While there is clearly a cyclic relationship, there is still a debate on whether the end goal should be unconscious competence (where the learner finds it difficult to explain what s/he is doing) or conscious competence, where the learner can explain what s/he is doing (Cheetham & Chivers 2005).

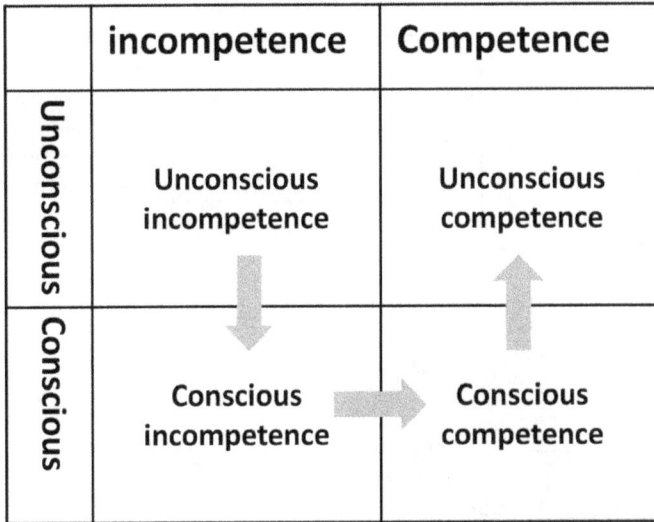

	incompetence	Competence
Unconscious	**Unconscious incompetence** ↓	**Unconscious competence** ↑
Conscious	**Conscious incompetence** →	**Conscious competence**

Figure 1: Four-quadrant representation of the conscious competence theory.

Individuals do not stay unconsciously competent all the time unless they continue practicing their skills and also keep making adjustments and improvements in their competence. If they do not do so, they may lose their competency. Supporting this hypothesis, a new body of arguments has emerged, which indicates that, in the 'unconscious competence' stage, individuals may cease to learn further because of which they may lack the knowledge or skills on newer methods, thus finding themselves once again in the 'consciously incompetent' stage. At this stage, however, individuals are aware of what they do not know (Figure 2).

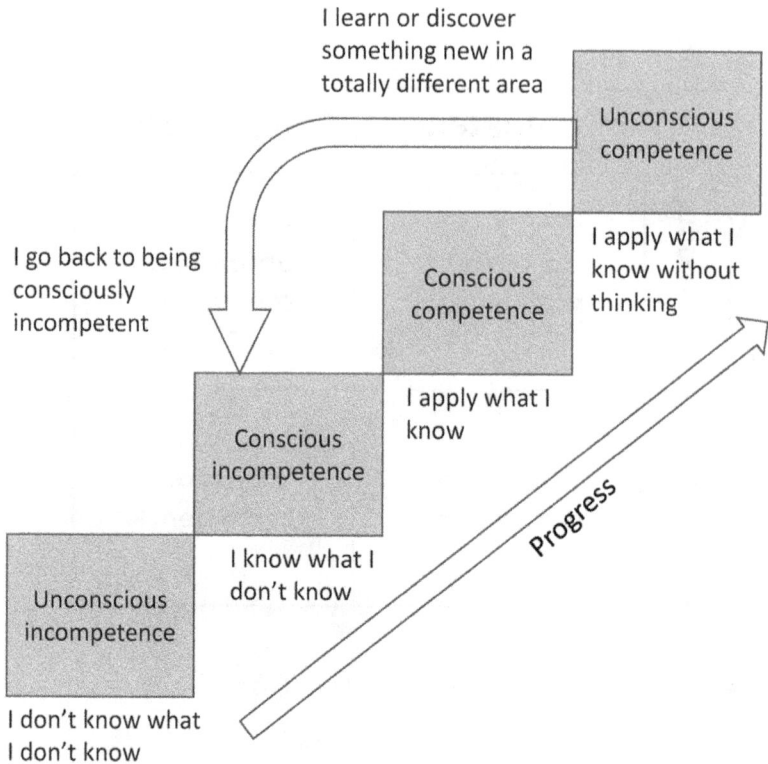

I learn or discover something new in a totally different area

Unconscious competence

I go back to being consciously incompetent

Conscious competence

I apply what I know without thinking

Conscious incompetence

I apply what I know

I know what I don't know

Progress

Unconscious incompetence

I don't know what I don't know

Figure 2: Skill acquisition, progression, and regression to previous stages (adapted from elijahconsulting.com).

To represent this loop, William Taylor in 2007 suggested a 'reflective competence model' (Business Balls, n.d.) with a fifth level called 'reflective competence' that represents a continuous self-observation to keep ideas and skills fresh, and also allows for further skill development (shown in Figure 3).

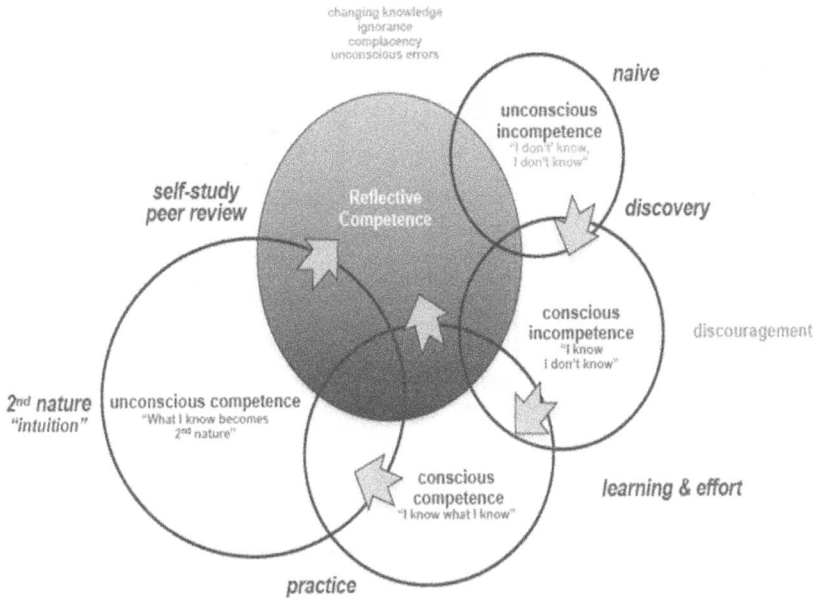

Figure 3: Reflective competence model (Taylor 2007 and Chapman, n.d.).

However, the reflective competence model, though very simple, does not explain the exact mechanisms of proficiency development. On the contrary, it is possible to apply the model to repetitive and familiar tasks where it is possible to achieve automaticity through practice. The reflective competence stage also suggests that learners need to be aware of what they need to know to tackle novel situations in which automaticity may not help them.

Another representation by Drejer (2000) included a superimposition of the Dreyfus and Dreyfus (1986) five-stage skill acquisition model, as shown in Figure 4. The Gordon Institute USA (www. gordontraining.com) is known to use this as a learning model in their training programs.

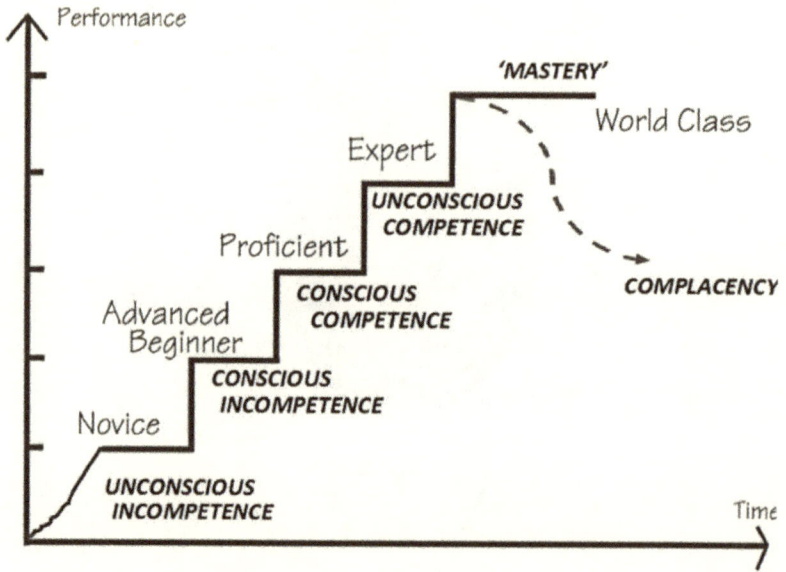

Figure 4: The Dreyfus & Dreyfus (1986) five-stage skill acquisition model based on the conscious competence theory (adapted from Drejer 2000).

The issue with the conscious competence models is that they conceptually explain the way in which individuals learn but do not provide a means to quantify or segregate their developmental paths. There is no clear-cut indication on how to measure the attainment of each stage or where a particular stage ends. Further, we do not know whether the models are grounded in research or not.

3.2   FITTS & POSNER (1967) MODEL OF SKILL ACQUISITION

Fitts (1964) suggested that the acquisition of perceptual-motor skills occurs in three stages, which supports the progression of an individual from a conscious to a less conscious form of practice. The three stages are:

- **Cognitive stage:** Learners consciously and constantly interact with nature and the mechanics of what is being done. In the cognitive stage, learners are still trying to understand the instructions and develop performance strategies. Cornford & Athanasou (1995, p. 12) indicate 'basic factual understandings, the broad outline, the essential nature of each of the steps and the order in which these must be performed.' Schneider (1993, p. 316) characterizes this stage in terms of the 'effort to understand the task demands and to distinguish between important and unimportant aspects of the task. The focus is on the acquisition of declarative knowledge about the task.' The stage may involve activities like reading, having discussions, or gathering information (Oyewole *et al.* 2011).

- **Practice fixation stage/Associative stage:** Fitts & Posner (1967) refined the original model by Fitts (1964) and changed the term *fixation* to *associative*. In this stage, repetition of skills and involvement with reality increase the depth of understanding. Here, the steps to attain and sequence of skilled performance get imprinted in permanent in memory of the learners. Performance strategies get refined at this stage. Prior learning is leveraged to develop strategies in new situations. Associations are developed by the learners between clues and responses for 'making the cognitive processes more efficient to allow rapid retrieval, thus transforming declarative knowledge into procedural forms' (Schneider 1993, p. 316).

- **Autonomous stage:** In this stage, learners execute their skills automatically or subconsciously. They become skilled in their tasks to a level where they do not require a lot of resources. More cognitive resources are available to process other activities. Learners use their conscious minds in monitoring and solving problems. In this stage, speed and accuracy are improved without having to change the approach and understanding of the domain

(Oyewole *et al.* 2011). However, the rate of improvement will hit a plateau after a certain amount of practice.

The last stage (autonomous stage) of this model highlights the 'subconscious' use of a skill (indicating automaticity), whereas the unconscious competence theory called this last stage as the 'unconscious' use of skills. However, Fitts' usage of the term "subconscious" appears to be more appropriate with respect to the underlying mechanism. Ericsson (2009a) believes that the Fitts & Posner (1967) model is helpful in achieving automaticity in everyday skills but is not in any way an expert performance model (p. 12).

The major implication of Fitts & Posner's model toward proficiency/expertise development is the laying out of the instructional activities in stages. The model describes the sequence of events proposed for the development of a skill, which then provides the mechanism and guidance for the instructional events by which automaticity is achieved. Clavarelli, Platte & Powers (2009) suggested instructional strategies as well as assessment strategies for learners at each stage of the Fitts & Posner model.

3.3   ANDERSON (1982) ACT-R THEORY OF COGNITION

Anderson (1982) proposed a three-stage learning model which described the progression in learning from declarative knowledge, through knowledge compilation to procedural knowledge. The Anderson model has been revised a few times, and the updated model is called the Adaptive control of thought-rational (ACT-R) theory of cognition (Anderson 1990; Anderson & Lebiere 1998), which is now considered a general theory of cognition.

The ACT-R model proposes three types of memory structure through which skills are learned: declarative memory, procedural memory, and working memory, as shown in Figure 5 (Anderson 2000).

- **Declarative memory:** According to the ACT-R theory, all knowledge in the form of facts and information first begins as declarative information that is stored in the declarative memory of the human brain. Declarative memory mainly links the propositions, images, and sequences by association.

- **Procedural memory:** Procedural knowledge is acquired by making inferences from the already existing factual knowledge in the declarative memory. Procedural memory, also called long-term memory, stores information in the form of representations, described by Anderson (1982) as "productions." Each production has a set of conditions and actions based in the declarative memory.

- **Working memory:** The nodes of long-term memory all have some degree of activation. Working memory is that part of the long-term memory that is the most activated.

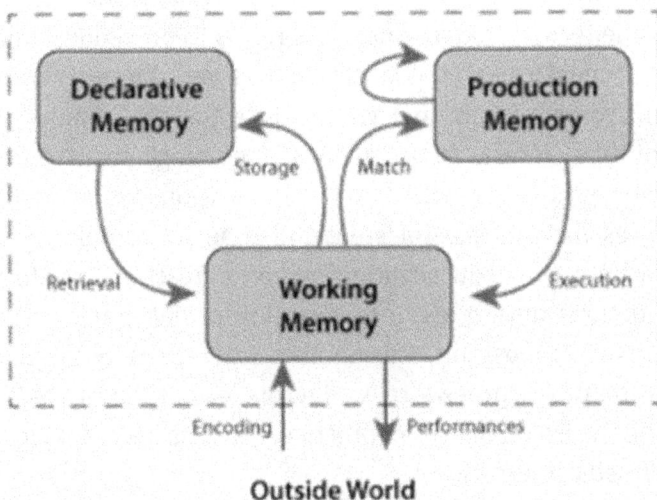

Figure 5: ACT-R model interactions (adapted from http://www.instructionaldesign.org/theories/act.html)

The ACT-R model supports three fundamental types of learning: (a) generalization, in which productions become broader in their range of application, (b) discrimination, in which productions become narrow in their range of application, and (c) strengthening, in which some productions are applied more often than the others. New productions are formed by the conjunction or disjunction of the existing productions.

During the acquisition of cognitively complex skills, learners pass through three stages of knowledge formation:

- **Declarative knowledge:** Knowledge relevant to the performance of skills begins in a declarative form. Declarative knowledge includes facts, processes, principles, concepts, and information, etc., that learners receive about the topic of skill.

- **Knowledge compilation:** With practice, the knowledge on skills is converted gradually from the declarative to new procedures that may be applied without much attention effort. The ACT-R theory uses the term "productions" that refers to procedural knowledge 'which consist [sic] of goal statements and the overt actions and cognitive operations that will achieve the goals under specified conditions' (Clark 2006, p. 162). This stage involves domain-specific proceduralization and composition. While composition involves making several steps into one by combining adjacent procedures, proceduralization involves creating domain-specific productions during the interpretation of declarative knowledge. Through composition, the number of steps is reduced on each iteration of the procedure. While this leads to improved learning during the initial phases of practice, learning becomes slower with increased composition.

- **Procedural knowledge:** Procedural knowledge refers to the representation of what needs to be done in a given situation. In this stage, practice refines the appropriate procedures. Responses become more generalized and automatic. The ACT-R theory also suggests that with use over time, productions become automatic and unconscious. Oyewole *et al.* (2011, p. 1987) explain that 'actions are based on the presence of some conditions. Procedural knowledge is incorporated with production rules–condition–action pairs associated with the presence of a particular data pattern in working memory with the performance of a particular action. When a condition of a production rule is satisfied, the production can then apply and the action will proceed.' Ge & Hardré (2010), in their review, cited Anderson (2000) suggesting practice as essential for developing learned procedures to reach automaticity of expertise in any field.

The ACT-R theory indicates that both declarative knowledge and procedural knowledge reside in separate long-term memory systems. Clark (2006, p. 163) clarifies the interplay of new problems to solve and use of declarative knowledge:

> When the goal is to learn or solve a problem, we apply the goal-directed productions we have available and if available productions are incomplete or inadequate, we use declarative knowledge to fill in the missing steps. If instruction provides the necessary steps to fill the gap, learning occurs faster and more effectively (e.g., with fewer performance errors) than if we must fill in the gaps using declarative knowledge.

From the expertise development standpoint, the major implication of the ACT-R theory is the cognitive process involved in the development of automaticity, a state of expertise. Oyewole *et al.* (2011, p. 1987) support that the ACT-R theory is 'incorporated with several learning mechanisms such as translating verbal instructions into procedures for performing a task, generalizing and differentiating existing procedures, composition and strengthening.' These mechanisms are translated into instructional strategies. As a result, the ACT-R theory has been used in several areas including learning, memory, problem-solving, decision-making, language, communication, perception, attention, cognitive development, and individual differences.

To apply the ACT-R theory in skill acquisition training, Clark (2010) suggested instructional techniques such as: goals and reasons (intentional module, goal buffer), overview (declarative module), connection to prior knowledge (declarative module), new knowledge (declarative, visual/manual modules), demonstrated procedures (production system, declarative module), practice (integration of new and old to achieve the goals), and feedback (intentional module to correct and perfect).

3.4   ACKERMAN (1988) THEORY OF ABILITY DETERMINANTS OF SKILL ACQUISITION

Ackerman (1988) presented an integrated theory that was partly based on an analysis of ability measures and partly on the nature of information processing and skill acquisition. The theory described the individual differences in 'ability structure' during the different phases of skill acquisition. Ackerman described that abilities were structured at different stages based on three components: complexity processing, the content of information (figural, verbal, numerical, content etc.), and *speededness* of processing. He described that an individual develops ability structures as he or she passes through the stages of skill acquisition. He reasoned that the ability–performance relationship is the function of three task characteristics, namely, consistency of information processing demands, task complexity, and degree of task practice. He proposed that, in tasks characterized by consistent information processing components, the performance of an individual in the cognitive phase is determined by general and broad content abilities. The performance in the associative phase is determined by the perceptual speed abilities. Lastly, the performance in the autonomous phase is determined by the psychomotor abilities of an individual. However, for inconsistent tasks, general and broad content abilities determine the performance across all three phases. Similar to tasks involving consistent information processing, here also perceptual speed abilities determined the performance in the associative phase while psychomotor abilities determined performance in autonomous phases.

On the skill front, Ackerman supports that skill acquisition follows phases similar to those proposed by Fitts & Posner (1967), and refined by Anderson (1982) and Shiffrin & Schneider (1977).

- **Cognitive phase:** Ackerman (1992) considers that 'at cognitive phase, there is a high cognitive load on the learner in the context of understanding task instructions, general familiarization with task goals, and formulating strategies for task accomplishment. Once the learner has acquired the basic procedures, he or she proceeds to the associative phase of skill acquisition' (Ackerman 1992, p. 599). He reasoned that to go beyond the cognitive stage, a learner needs to first develop consistent information processing characteristics.

- **Associative phase:** Ackerman opined that the associated phase of skill acquisition involves 'the proceduralization of task strategies in a manner that makes performance quicker and less error prone' (Ackerman 1992, p. 599).

- **Autonomous stage:** According to Ackerman, the final stage of skill acquisition is the autonomous stage, which 'involves the automatization of task skills, such that performance of the task, once engaged by the learner, often proceeds with little or no attentional effort. Performance at this phase of skill acquisition is typically highly speeded [sic] and highly accurate' (Ackerman 1992, p. 599).

From the expertise building standpoint, the major implication of this model is that stages 1 and 2 require cognitive resources. The need for cognitive resources slows down both the speed and performance, a phenomenon that occurs predominantly in the case of novel, complex, and inconsistent tasks. Therefore, it may be inferred that the sequence of tasks based on task characteristics may be an important determinant of expertise development.

Subsequent research studies raised a few flaws in this model. For instance, Langan-Fox *et al.* (2002, pp. 104–105) opined that Ackerman's theory 'largely ignore [*sic*] the experiences and internal processes of a person (e.g., how they feel, what strategies they are using, the role of the external environment).'

3.5 VANLEHN (1996) COGNITIVE SKILL ACQUISITION THEORY

Developing upon the three-stage simplified theory by Fitts (1964), VanLehn (1996) reviewed the research on cognitive skill acquisition and explained the findings in the form of three stages of cognitive skill acquisition. Though the purpose of this model was not to explain learning, several researchers (e.g., Kim, Ritter & Koubek 2013) appear to view the model as a simplified consolidation of the various phenomena and methods employed at the different stages of the cognitive skill acquisition:

- **Early-stage:** This stage is characterized by learners trying to understand the domain knowledge without applying it.

- **Intermediate stage:** When learners enter this stage, they have some background information on solving the problems. Their knowledge is not complete and may be missing or incorrect. At this stage, learners start acquiring heuristics and experiential knowledge for solving problems. They prefer to understand the problems through examples (Carroll 1994; Cooper & Sweller 1987). Toward the end of this stage, learners become almost flawless, and possess the ability to solve problems without conceptual errors.

- **Late stage:** In this stage, the speed and accuracy of problem-solving of the learners are enhanced as they start practicing more and more. However, the basic approach to solving problems does not change. This is the stage at which the power law of practice holds good. The primary learning mechanism is the one described by Anderson (1982) that knowledge is converted (or compiled) from declarative knowledge to procedural knowledge. With practice, the speed of applying some pieces of procedural knowledge increases (Carlson *et al.* 1990), and several parts of the cognitive skills become automatic.

3.6 KIM *ET AL.* (2013) THEORY OF SKILL RETENTION

Based on the Anderson (1982) ACT-R theory of cognition, and using Fitts (1964), Rasmussen, Pejtersen & Goodstein (1994), and VanLehn (1996), Kim, Ritter & Koubek (2013) suggested a new learning theory that integrates learning and the impact of forgetting during the declarative, mixed, and procedural stages:

- **Declarative stage:** According to the ACT-R model, both learning and forgetting happen at this stage, which dictates that an item from declarative memory is activated depending upon how often and how recently it is accessed (Anderson, 1982). Practice allows the generation of task-specific procedural knowledge as opposed to the need for the retrieval of declarative string. Kim, Ritter & Koubek (2013) postulated that declarative memory degrades quickly if one does not use the knowledge or practice it. Learners may reach a point where retrieval from memory may fail, or the response time may increase. Such memory load will lead to degraded performance.

- **Mixed stage:** In this stage, task knowledge is represented by a combination of declarative knowledge and procedural memory. Thus, the mix of declarative memories and procedural memories is used to learn the task. The mixture varies based on the tasks. When learners do not use declarative knowledge, it leads to forgetting it, which eventually ends up as mistakes, especially in the case of low-frequency tasks. Since declarative memory may not be fully activated in this stage, training is required to keep it active in order to generate new procedural rules. However, procedural memory does not appear to be decay at this stage.

- **Procedural stage:** In this stage, task knowledge is available in both declarative and procedural forms, but performance is mainly governed by the procedural knowledge. Practice compiles task knowledge as 'proceduralised skill' (Kim, Ritter & Koubek 2013, p. 27). Though lack of practice may decrease declarative knowledge, learners can still perform the tasks if all the knowledge is proceduralized.

While this model is the superimposition of the Anderson (1982) ACT-R model on the Fitts & Posner (1967) model, from a performance perspective, it does explain the link between procedural knowledge and declarative knowledge. Hence, to develop performance, learners may need spaced reinforcement training and continuous practice.

3.7 VYGOTSKY (1978) ZONE OF PROXIMAL DEVELOPMENT

Vygotsky (1978), on the basis of developmental psychology, came up with a theoretical framework of the development of cognition based on the studies with children. His framework explains learning within the individual's immediate context or environment. His fundamental premise is that social interactions play a more significant and essential role in the development of cognition. He further argued that cognitive development depends on the "zone of proximal development (ZPD)," which he defined as a level of development that is attained when one engages in social behavior. Vygotsky (1978, p. 78) defines ZPD as 'the distance between the actual developmental level as determined by independent problem solving and the level of potential development as determined through problem solving under adult guidance, or in collaboration with more capable peers.' Full cognitive development depends on total social interactions. The range of skills that can be developed with social and adult interactions far more exceeds that achieved alone (Billett 2000).

From the expertise development standpoint, Bransford *et al.* (2004, p. 81) stress that 'The zone of proximal development embodies a concept of readiness to learn that emphasizes upper levels of competence.' Another implication of this model is that proficiency grows when one learns in context. Combining the contextual and social aspects of learning, Lave and Wagner developed the concept of "situated learning." This framework by Vygotsky (1978) also suggests the key role of capable peers, colleagues, and people around the learners, indicating the role of informal learning alongside formal learning. Underlying to this is what may manifest as mentoring and guided learning (Billett 2000). The Vygotsky (1978) framework shares some similarities with models emphasizing the social aspect of learning and proficiency development, such as the Bandura Social Learning Theory, Lave and Wegner Situated Learning Theory, and Collins Cognitive Apprenticeship Model.

Dunphy & Williamson (2004) suggest that learning in the ZPD occurs in four stages.

- **Assisted phase:** In this phase, performance is assisted by more capable peers. Here, since learners have only limited understanding, competent peers or experts provide direction and modeling. In the beginning, learners may be imitating their peers but soon develop an understanding of the overall performance.

Further assistance is given to the learners through feedback, questions, or other means like providing task or subtask-specific goals. In this phase, the learners start taking up more responsibilities and also start demonstrating self-regulation. Dunphy & Williamson (2004, p. 116) suggest that the task of this stage is 'accomplished when the responsibility for tailoring the assistance, tailoring the transfer and performing the task itself has been effectively handed over to the learner.' The primary task of experts in this stage is to provide accurate and personalized assistance to the learners and respond to their efforts.

- **Self-assistance phase:** In this stage, learning is self-guided, and learners perform the tasks with no assistance. However, performance is not fully automatic yet.

- **Automated performance phase:** In this phase, performance is fully developed and automated. Learners no longer need any assistance from experts. Tasks become internalized, and the need for self-regulation vanishes. This is the point that marks the end of ZPD, and assistance at this stage may become a performance degrader and may become disruptive to the learners.

- **De-automatization of performance phase:** It is observed that 'under certain circumstances practitioners may drop to a lower level (phase) of performance' (Dunphy & Williamson 2004, p. 123). De-automatization of performance brings down learners back through ZPD again. During this phase, the goal is enhancement, improvement, and maintenance of performance. Recurrent cycles of assistance both from self and others are likely to occur. Dunphy & Williamson (2004, p. 116) suggest:

  > 'De-automatization and recursion occur so regularly that they constitute a stage four of the normal developmental process. What one formerly could do, one can no longer do. After de-automatization, for whatever

reason (environmental changes, stress, major upheavals, trauma), if capacity is to be restored, then the developmental process must become recursive.'

This model has several similarities with the models by Fitts & Posner (1967) and Ackerman (1988) with respect to its first three stages, and with the conscious competence theory with respect to its last stage. From an expertise standpoint, Dunphy & Williamson (2004, p. 121) opined that 'the ZPD contains a clear outline of how individual skills can be acquired. The ZPD is the only model to clearly identify that under certain circumstances, the performance of skills may deteriorate, and also outline processes by which these skills may be regained (recursion through a lower level within the ZPD) [sic].' However, this model also has its limitations. There is not much literature on ZPD that provides details on how learners transition from the autonomous performance stage to the expertise stage (Dunphy & Williamson 2004).

3.8   DREYFUS & DREYFUS (1986) STAGES OF SKILL ACQUISITION

The most recognized work among all in specifying the stages of proficiency was that proposed by Dreyfus & Dreyfus (1980, 1986, 2008, 2009), supplemented by Dreyfus (2005, 2004). 'Dreyfus model has been [sic] its derivation from observation of the performance of experts, such as jet pilots and dancers, experts who are used to tackling direct problems' (Peña 2010). The Dreyfus & Dreyfus model is based on the notion that skill acquisition is a continuous process and that skills are transformed into performance by experience and mastery, which then brings a change in performance. They believe that experience with real-time situations alone produces higher levels of performance.

Dreyfus & Dreyfus (1986) proposed that during the process of skill acquisition, learners pass through five distinct levels of proficiency:

novice, advanced beginner, competent, proficient, and expert. In their original model (Dreyfus & Dreyfus 1980), they did not have the stage named "advanced beginner." Each stage of the Dreyfus & Dreyfus model of skill acquisition involves qualitatively different perceptions of skills and/or the mode of decision-making. According to Dreyfus and Dreyfus (1986), the most important difference between the different levels of expertise is the gradual shift from using analysis skills to intuition and the level of involvement. As a novice progresses, s/he acquires more and more situational understanding and is able to exert his intuition in several different situations. Dreyfus & Dreyfus (2005, pp. 782–788) explained their model with the help of a simple example of a car driver:

> The student automobile driver learns to recognize such domain-independent features as speed (indicated by the speedometer) and is given rules such as shift to second when the speedometer needle points to 10....The advanced beginner driver uses (situational) engine sounds as well as (non-situational) speed in deciding when to shift. He learns the maxim: 'Shift up when the motor sounds like it's racing and down when it sounds like it's straining.' Engine sounds cannot be adequately captured by a list of features, so features cannot take the place of a few choice examples in learning the relevant distinctions.....A competent driver leaving the freeway on an off-ramp curve learns to pay attention to the speed of the car, not whether to shift gears. After taking into account speed, surface condition, criticality of time, etc., the driver may decide he is going too fast. He then has to decide whether to let up on the accelerator, remove his foot altogether, or step on the brake, and precisely when to perform any of these actions. He is relieved if he gets through the curve without mishap, and shaken if he begins to go into a skid....The proficient driver,

approaching a curve on a rainy day, may *feel in the seat of his pants* that he is going dangerously fast. He must then *decide* whether to apply the brakes or merely to reduce pressure by some specific amount on the accelerator. Valuable time may be lost while making a decision, but the proficient driver is certainly more likely to negotiate the curve safely than the competent driver who spends additional time *considering* the speed, angle of bank, and felt gravitational forces, in order to *decide* whether the car's speed is excessive....The expert driver not only feels in the seat of his pants when speed is the issue; he knows how to perform the appropriate action without calculating and comparing alternatives. On the off-ramp, his foot simply lifts off the accelerator and applies the appropriate pressure to the brake. What must be done, simply is done.

The following is a general description of the five stages:

Stage 1 – Novice: In this first stage, learners (novices) work to gain a better understanding of their skills mostly through formal training. However, the learners continue to be unaware of the specific skills or knowledge that must be applied in real-world situations; they indicate an interest and willingness to develop the necessary skills and knowledge. During this phase, novices learn to recognize the various facts and figures pertaining to the desired skills and the rules to apply them. Novices view these facts and figures as context-free while applying these skills. They are trained to adhere to the rules rigidly and apply them based on the situation. However, from that perspective, novices do not possess high situational perception or discretionary judgment to decide whether a given rule is to be applied or not in that situation.

Stage 2 – Advanced beginner: Novices move up to the next stage of advanced beginners as they start gaining experience in real-world situations, and their performance starts improving to a marginally acceptable level (DiBello, Lehman & Missildine 2010). At this stage, learners start comprehending objective facts, initial concepts, and specific rules and start applying them within specific disciplines or in structured settings but may struggle to apply them to real-world situations. As novices gain more practical and concrete experience, they start comparing the new situations with the situations faced previously. However, they can only apply the earlier-learned rules with the help of which they deal with unrecognized facts and elements. At this stage, the learners start applying more sophisticated rules to both context-free and situation factors. While these rules make it possible for the advanced beginners to learn from their experience, their situational perception is still limited. Learners may be comfortable solving routine, well-defined problems but may be ineffective and inefficient in applying their knowledge in unfamiliar settings or in solving ill-defined problems.

Stage 3 – Competent: With experience, learners begin to recognize more and more context-free and situational elements. At this stage, learners are able to recognize the situations and concentrate on the relevant important elements. They are able to assess situations, set goals, and choose the best course of action. While they start handling situations, they may or may not be able to fully apply the rules. Thus, they may or may not be successful, but that constitutes an important element of future expertise.

Stage 4 – Proficient: At this stage, the learners are deeply involved in their tasks. They are capable of identifying the important elements or components of the tasks and paying the requisite attention. Proficient learners start viewing situations holistically in terms of various elements. As situations change, their deliberation, plans, and assessments also change. With changing situations, they start seeing new patterns that are different from normal behaviors. The decision-making is very quick and fluid because of their experience in a similar situation in the past. Still, proficient learners use maxims to guide their decision-making. Consistency in performance distinguishes this phase from the 'competent' stage. Benner (2004, p. 198), in her nine studies spanning over 21 years, observed that 'The most qualitatively distinct difference lies between the competent and proficient level [*sic*], where the practitioner begins to read the situation. The proficient performer begins to increasingly change his or her perception of the nature of the situation and then deliberates about changing plans or strategies in response to the new understanding of the situation.'

Stage 5 – Expert: Experts do not apply rules, or use maxims or guidelines. Thus, in this stage, learners have an intuitive grasp of the situations based on their deep tacit understanding. One of the key aspects of this stage is that learners rely on intuition, and an analytical approach is used only in case of new situations or unrecognized problems that were not experienced earlier. Experience-based deep understanding enables the learners to display a fluid performance. At this stage, skills become so automatic that even the learners (experts) are not aware of them. Based on prior experience, they can even come up with solutions for situations that are newer and have never experienced before (DiBello, Lehman & Missildine 2010). Experts adopt a contextual approach to problem-solving and start understanding the relative, non-absolute nature of knowledge. This key ability distinguishes the "expert" practitioner from the "proficient" practitioner (*D'Youville College, n.d.*). Reflection comes naturally, and experts solve problems almost unconsciously.

To date, the Dreyfus and Dreyfus model has remained the most simplistic and most commonly used model of stages of skill progression due to its implications in the professional and training world. Over the years, several researchers such as Gunderman (2009), Benner (2004), Eraut (1994), Khan & Ramachandran (2012), and Dreyfus & Dreyfus (1986, 2004, 2005, 2008, 2009) have significantly added more perspectives to the definition of each of the levels/stages.

Eraut (1994) summarizes the stages of the Dreyfus and Dreyfus model as follows (Table 2).

Table 2 Stages of the Dreyfus and Dreyfus model, summarized by Eraut (1994)

Stage	Characteristics
1. Novice	Rigid adherence to taught rules or plans Little situational perception No discretionary judgment
2. Advanced beginner	Guidelines for action based on attributes or aspects (aspects are global characteristics of situations recognizable only after some prior experience) Situational perception is still limited All attributes and aspects are treated separately and given equal importance
3. Competent	Coping with crowdedness Now sees actions at least partially in terms of longer-term goals Conscious, deliberate planning Standardized and routinized procedures
4. Proficient	Sees situations holistically rather than in terms of aspects Sees what is most important in a situation Perceives deviations from the normal pattern Decision-making less labored Uses maxims for guidance, whose meanings vary according to the situation
5. Expert	No longer relies on rules, guidelines or maxims Intuitive grasp of situations based on the deep tacit understanding

Analytical approaches used only in novel
situations or when problems occur
The vision of what is possible

However, based on the in-depth interviews with the Dreyfus brothers, Flyvbjerg (1990, 1991) contested that the Dreyfus & Dreyfus model did not account for progressive innovation and practical wisdom. Subsequent work by Dreyfus and Dreyfus (2001) includes a sixth stage called 'mastery,' beyond the level of expertise.

Stage 6 – Mastery: Dreyfus & Dreyfus (2008, p. 120) differentiated mastery from competence or expertise by stating that 'A very different sort of deliberation from that of a rule-using competent performer or of a deliberating expert.' Khan & Ramachandran (2012, p. 5) summarize the characteristics of a learner in this stage as 'Performance becomes a reflex in most common situations. Sets new standards of performance. Mostly deals with complex situations intuitively. Has a unique vision of what may be possible related to the given task. Able to train other experts at national or international level.'

Stage 7– Practical Wisdom: Subsequently, Dreyfus & Dreyfus (2008) added the seventh stage called 'practical wisdom' in their original model of skill acquisition.

Chi, Glaser & Farr (1988) reported the strength of this model as 'the case it makes for tacit knowledge and intuition as critical features of professional expertise in unstructured problem areas' (Scobey 2006, p. 163). The Dreyfus and Dreyfus model has proven to be extremely useful in depicting the levels of expertise in any profession. Numerous attempts have been made to apply the model to several professions such as clinical practice (Peña 2010), healthcare (Greene, Lemieux & McGregor 1993; McElroy, Greiner & de Chesnay 1991; Khan &

Ramachandran 2012); correctional services (Scobey 2006); education (Bedi 2003); and nursing (Benner 1984, 2001, 2004; Ramsburg 2010).

The most renowned among these studies is the one by Benner (2004), who conducted nine studies over a period of 21 years applying the Dreyfus and Dreyfus model. She showed that as the clinical nurses advance in their experience, they become more and more proficient in the clinical environment. Benner (2004, p. 198) found that the Dreyfus and Dreyfus model was 'predictive and descriptive of distinct stages of skill acquisition in nursing practice.'

From a workforce development standpoint, organizations like to most of their employees to 'proficient' stage while there may be several of them coming up to speed at 'advanced beginner' and 'competent' levels. However, some of the employees typically get intuitive skills developed to a level where they are considered as operating at an 'expert' level.

Literature shows evidence of the application of the Dreyfus and Dreyfus model to complex hi-tech jobs such as software engineering (Page-Jones 1998). Based on these studies, it has been inferred that the model can be applied to other hi-tech industries such as aviation, engineering, manufacturing, and scientific research to explain how an employee's ability to solve complex problems increases progressively as s/he moves up toward the expertise stage. Despite the widespread simplicity and appeal of the Dreyfus and Dreyfus model, another school of thought questioned its validity. For example, Day (2002) noted that it was difficult to apply five stages in professional settings because practitioners perform a range of tasks in their jobs, and they will not fit into one stage for all the tasks they do. Thus, the most common objection raised was the representation that proficiency acquisition is a linear process independent of the influence of external factors and domain expertise (Grenier & Kehrhahn 2008). Another objection to this model was 'its failure to explain how someone becomes an expert and its stress on the importance of experience and not of its impact' (Farrington-Darby & Wilson 2006, p. 29). Other concerns included the absence of a social structure in the process of knowledge and skill acquisition, lack of objective quantification on how to measure the attainment of each stage or where a particular stage ends, and the lack of operational definitions of intuition (Peña 2010). Even though these staged skill acquisition models conceptually explain how an individual learns, 'one cannot accurately predict where people are in the skill-acquisition process' (Langan-Fox *et al.* 2002, p. 106).

Though the Dreyfus and Dreyfus model does not directly lend itself to defining instructional strategies in the context of proficiency and expertise development, some researchers have proposed a few instructional models based on it. Ross *et al.* (2005) created a comprehensive set of training guidelines on tactical decision-making based on the Dreyfus and Dreyfus model (Clavarelli, Platte & Powers 2009, p. 3):

> '*Novice:* Give learners the rules needed to guide performance, provide direct coaching and close mentoring on tasks and strategies. *Advanced Beginner:* Emphasize experiential learning through scenarios designed to illustrate recurring patterns and instances requiring the learner to formulate his/her own strategies and task performance guidelines. *Competent:* Increase complexity and variety of task performance conditions and full mission scenarios to include higher-level planning and decision making. *Proficient:* Require learner to formulate and test probable solutions given exposure to novel situations. Increase mission/task complexity and time pressure. *Expert:* Insert lessons from real-world case history and operational experience. Train with and against other expert level performers. Conduct peer evaluations and feedback.' (Ross *et al.* 2005, p. 67)

3.9   ALEXANDER (1997) MODEL OF DOMAIN LEARNING

Alexander (1997) introduced the 'model of domain learning' (MDL) that focused on the journey of a learner, 'conceptualized as systematic changes within and across stages of development' (Alexander 2003a, p. 10). Alexander (1997) postulated that the stages of expertise development were governed by the nature of the interplay of three components, namely, knowledge (domain and topic), strategic processing (surface and deep), and interest (long-term and situational).

Alexander (2003a) suggests two kinds of strategies that learners employ during their progression: surface-level and deep-processing. According to Alexander (2003a), 'Surface-level strategies (e.g., rereading or paraphrasing) are processes individuals use to make sense of the text. Deep-processing strategies, by comparison, involve delving into that text, as when students judge author credibility or form mental representations' (p. 11.). While surface-level strategies are used by learners to understand unfamiliar content, deep-processing strategies allow them to follow an analytical approach about the topic to be learned. However, Alexander (2003a) asserts that the deep-processing strategies that she refers to in her model are different from the proceduralization or consolidation of knowledge suggested by Anderson (1981, 1982) in the ACT-R model.

Further, the Alexander (1997) MDL model also specifies the two types of interests that are considered in expertise development: individual and situational.

Individual interest is the enduring interest by the learners to stay engaged with the experiences. Individual interest may be of two types: general or professional (VanSledright & Alexander 2002). Professional interest is more in line with vocational activities.

Situational interest is the interest triggered by immediate situations or events and is considered key to expertise development.

This model presents the journey toward proficiency or expertise as a three-stage process. These stages are acclimation, competence, and proficiency/expertise (shown in Table 5):

- **Acclimation:** In this stage, learners possess only limited and fragmented knowledge, and they are yet to form a 'well-integrated body of domain knowledge' (Alexander 2003a, p. 11). Therefore, their ability to discern the situational elements is limited. Most of the problems encountered by the learners during this stage are new and challenging, which trigger a surface-level strategic process. Thus, situational interest is very important to keep learners focused on the topic.

- **Competence:** Alexander (2003a, p. 12) stated that 'competent individuals not only demonstrate a foundational body of domain knowledge, but that knowledge is also more cohesive and principled in structure.' Since the problems at this stage are somewhat familiar, learners apply a mix of surface-level and deep-processing strategies. Also, the individual personal interest in the problems is now much more marked than that in the acclimation stage.

- **Proficiency/Expertise:** Based on Alexander's model, Kim (2015, p. 4) summarizes the 'proficiency stage' as follows: 'In this stage, proficient learners, with increasing experience, construct sufficient contextual information and organize their knowledge base for a problem situation that includes some domain-specific principles. They conceptualize a sufficient problem space in all three dimensions that accommodate the real features of a given problem situation. The probability of resolving a problem markedly increases. Proficient learners sometimes represent a relatively small but efficient knowledge structure in which

49

sufficient key concepts are well organized (good structural and semantic dimensions with a surface dimension that is lacking).' The knowledge base of learners in this stage is both broad and deep. Alexander (2003a, p. 12) believes that learners who have reached this stage pose questions, conduct investigations, and push their boundaries in an attempt to 'problem finding.' This stage is marked by the use of deep-processing strategies and a very high level of personal interest in the topic.

The characteristics of these three stages, as suggested by Alexander (1997), are summarized in Table 3.

Table 3: Stages of proficiency development in Alexander (1997) MDL

Stage	Knowledge (domain/topic)	Strategic processing (surface/ deep)	Interest (individual or situational)
Acclimation	Learners have limited or fragmented domain and topic knowledge	Challenging tasks prompt the use of surface-level strategic processing	Reliance on situational interest to maintain learner focus and performance
Competence	Learners demonstrate the foundation body of knowledge	Use surface-level strategies and develop deep-processing strategies to acquire knowledge	Individual interest reduced the reliance on situational interest
Proficiency/ Expertise	Broad and deep topic/domain knowledge base	Use deep-processing strategies almost exclusively	High individual interest and engagement

Alexander (2003a, p. 11) states that 'the components of knowledge, strategic processing, and interest configure differently as individuals progress from acclimation to competence and proficiency/expertise.' Synergy among these components is necessary for the learner to move from competence into the proficiency /expertise stage.

Alexander (2003a, p. 11) has cited the examples of several studies which have applied the MDL model in different domains as a means to identify strategies toward expertise (e.g., Alexander, Jetton, & Kulikowich, 1995; Alexander, Murphy, Woods, Duhon, & Parker, 1997; Alexander, Sperl, Buehl, Fives, & Chiu, 2002; Murphy & Alexander, 2002; Chen, Shen, Scrabis, & Tolley, 2002; Lawless & Kulikowich, 1998; VanSledright, 2002). Very recently, Kim (2015, p. 12) noted in a study of 133 students in problem-solving that 'the study suggests that the three-stage model provides a better framework for measuring learning progress in complex problem solving.'

From the proficiency/expertise standpoint, this model has some useful implications. Baker (2006) cited Alexander (2003a, 2003b), claiming that suitable training interventions can be designed to move learners to the proficiency /expertise stage. This model is majorly used in managing the progress of learners and providing them with a practical instructional environment.

3.10 HOFFMAN (1998) PROFICIENCY SCALE

Hoffman (1998) recognized the challenges in defining expertise. He commented: 'If one acknowledges that expertise develops, and that qualitative changes occur over the developmental period, then one must make some attempt at stage-like categorization, if only to motivate research' (Hoffman 1998, p. 84). That is, if one can define novice and experts in terms of characteristics, knowledge, and performance, the remaining stages in the continuum of development can be retrieved from the literature. Based on the Craft Guilds in the

Middle Ages of Europe, Hoffman proposed that a novice develops into a master through 7 stages, similar to a craftsman, as shown in Table 4.

Table 4: Proficiency scaling proposed by Hoffman (1998)

Stage	Characteristics
Naïve	One who is ignorant of the domain
Novice	One who is new—a probationary member who has had some ("minimal") exposure to the domain
Initiate	One who has been through an initiation ceremony—a novice who has just started the profession
Apprentice	One who is learning—a student undergoing a program of instruction beyond the introductory level. Traditionally, the apprentice is immersed in the domain by living with and assisting someone at a higher level. The length of an apprenticeship depends on the domain, ranging from about one to 12 years in the craft guilds.
Journeyman	A person who can perform a day's labor unsupervised, although working under orders. An experienced and reliable worker, or one who has achieved a level of competence. It is possible to remain at this level for life.
Expert	The distinguished or brilliant journeyman, highly regarded by peers, whose judgments are uncommonly accurate and reliable, whose performance shows consummate skill and economy of effort, and who can deal effectively with certain types of rare or "tough" cases. Also, an expert is one who has special skills or knowledge derived from extensive experience with subdomains.
Master	Traditionally, a master is any journeyman or expert who is also qualified to teach those at a lower level. A master is a member of an elite group of experts whose judgments establish regulations, standards, or ideals.

> Also, a master can be that expert who is regarded by other experts as being "the" expert, or the "real" expert, especially with regard to subdomain.

The model has characteristics similar to the Dreyfus & Dreyfus (1986) model in which the apprentice corresponds to the advanced beginner while the journeyman corresponds to either the competent or proficient. Hoffman *et al.* (2014) opined that proficiency scaling should be one of the fundamental actions that need to happen in accelerated proficiency studies 'to forge a domain—and organizationally appropriate scale for distinguishing levels of proficiency' (p. 25). Explaining the suitability of changing knowledge and skill in progression, Macmillan (2015, p. 36) stated:

> Hoffman's Scheme, on the other hand, presumes a progression of knowledge and capabilities associated with different amounts of instruction and domain-specific experience. It describes relative levels of expertise or proficiency within a single knowledge domain. In doing so, it fulfils its traditional function as a means of describing the progression of increasing knowledge and skills as one moves, over many years, from the status of novice to a master in a specific field.

However, some researchers criticize the lack of scientific rigor of the framework. For example, Farrington-Darby & Wilson (2006, p. 29) commented, 'What this classification does not provide is the process that has to occur to move between the classifications.' Hoffman (1998) himself admitted that he did not create the model on the basis of a scientific principle.

The major implication of this scaling was that it offered a full spectrum of proficiency and suggested that individuals at different stages of their career may possess different levels of proficiency or competence (Hoffman 1998). Thus, this scaling taxonomy did not propose proficiency as a specific stage as suggested by the Dreyfus and Dreyfus (1986, 2005, 2006) model. Rather, it views that even a competent person has a certain level of proficiency in skills, tasks, or job functions, though qualitatively and quantitatively (if it can be measured) may be less than that of a proficient performer.

In organizations, we expect some employees to be at 'journeyman' stage while there may be some individuals who are still coming up to speed at the 'apprentice' level. Then some selected individuals may be considered 'experts.'

3.11   JACOBS (1997) LEVELS OF HUMAN COMPETENCE

Similar to the model proposed by Hoffman (1998), Jacobs (1997) proposed a 'taxonomy of human competence,' which suggested that human competence could be scaled with designations such as novice, specialist, experienced specialist, expert, and master, as shown in Table 5. The distinction was based on an individual's ability to produce outcomes. At the same time, it was recognized that '*master, expert, a specialist,* or a *novice* are usually relative notions' (Jacobs 2003, p. 7). While novices are those whose 'outcomes are less valuable or who produce no outcomes can have lower of competence [*sic*],' experts are those who 'achieve the most valuable outcomes in organisations.' Masters are considered the 'experts of the highest order' (Jacobs 2003, p. 5). Based on the value individuals produce in a given job role, Jacobs' taxonomy approached the stages of expertise development from a human resource development perspective, which included perspectives such as career progression and skill development.

Table 5: Levels of human competence proposed by Jacobs (1997)

Stage	Characteristics
Novice	One who is new to the work situation; minimal exposure to work; lacks the knowledge and skills necessary to meet the requirements set to adequately perform the work
Specialist	One who can reliably perform specific units of work unsupervised; the range of work is limited to the routine ones; however, needs coaching to use the most appropriate behaviors
Experienced specialist	One who can perform specific units of work; performs the work repeatedly; skillful and can perform the work with ease
Expert	One who has the knowledge and experience to meet and exceed the requirements of work; highly regarded by peers for their skills or expertise; can use skills for routine and nonroutine cases with minimum effort
Master	"The" expert among experts; the one, among the elite group, whose judgments are looked upon to set the standards

Both Hoffman (1998) in his proficiency scaling and Jacobs (1997) continued to position experts at the highest level of competence even though they specified master as the last stage. Jacobs (2003, p. 5) defines experts as 'people who possess the highest levels of competence are called experts. By definition, experts achieve the most valuable outcomes in organizations.' Only a few in an organization would qualify to be called as master as they are 'the experts of the highest order, and not all experts achieve this level of competence' (Jacobs 2003, p. 7).

Typically, organizations position development of most of their employees at 'experienced specialist' stage while there may be some individuals who are still coming up to speed at 'specialist.' Some brilliantly skillful individuals may be deemed as 'expert' level.

The definitions of novice, expert, and master presented by the taxonomy of human competence (Jacobs & Hawley 2002; Jacobs 2001, 1997, 2003; Jacobs & Washington 2003) are quite close to those presented by Hoffman (1998); however, the definitions of *specialist* level and *experienced specialist* presented by Jacobs' taxonomy align closely with those of the *apprentice* and *journeyman* levels presented by Hoffman (1998). Conceptually, this model has the same implications as Hoffman (1998) but has more organizational appeal with the Jacobs' (2003) structured on-the-job training (S-OJT) methodology.

3.12   ROSENBERG (2012) NOVICE-TO-EXPERT TRANSITION

Based on his experience, Rosenberg (2012) postulated that skill acquisition or development in learners is a four-stage process, each characterized by the way the learners/individuals perform the job in each stage of progression. These four stages are novice, competent, experienced, master/expert, as described below and shown in Figure 6:

- **Novice.** A novice (or apprentice) is, by definition, new to a job. Novices know nothing or only little about the work, certainly too little to be able to perform the job to an acceptable standard. Novices must be taught (or shown) the basics of the job before they can be expected to be productive. The learning strategy here is overwhelmingly instructional. "Show me (teach me) how to do my job," they ask.

- **Competent.** Competent (or journeyman) workers can perform jobs and tasks that meet the basic standards. They are those who have had their basic training and continue to look for more coaching and practice to get better at what they do. "Help me do it better," is their primary request.

- **Experienced.** This is where it gets really interesting. Experienced workers are beyond being merely competent. They can vary their performance based on situations. Since they are likely to encounter a variable and often unpredictable set of work problems and challenges, they need access to knowledge and performance resources on-demand, and the ability to search those resources in ways that are flexible and customizable by them, depending on the situation. "Help me find what I need," they ask, as they search for information, from sophisticated online systems to the coworkers around them.

- **Master/Expert.** Masters and experts create new knowledge. They invent new and better ways to do a job, and also are capable of teaching others to do it. They are truly unique individuals and seek to learn in a unique and personal way, primarily through collaboration, research, and problem-solving. "I'll create my own learning," they say.

The model borrows the nomenclature of the stages from the Hoffman (1998), Dreyfus and Dreyfus (1986), and Jacobs (1997) models. From a workforce perspective, organizations would position the development of most of their employees at the 'experienced' stage. At the same time, there may be some individuals who are still coming up to speed at a 'competent' level. There may be some truly exceptional employees who are considered to be at 'expert/master' level.

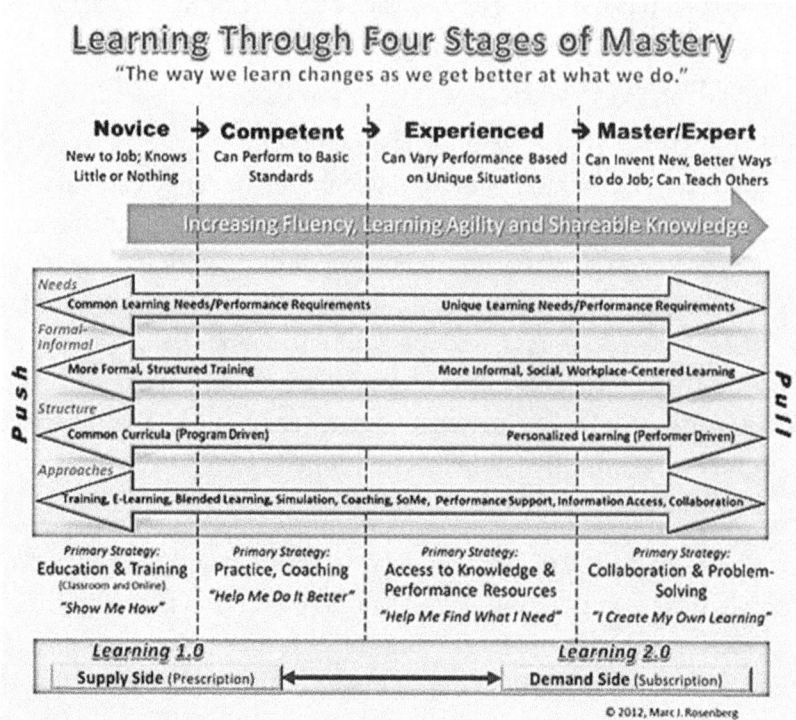

Figure 6: Learning through the four stages of mastery according to the Rosenberg (2012) model (adapted from Marc J. Rosenberg @ Marc My Words).

The Rosenberg (2012) model is a simple representation of how learning and job performance are integrated, and acts as a good guide to develop training for the appropriate audience. The model provides a framework for designers to put together programs appropriate for learners at different stages.

♣ ♣ ♣ ♣

CHAPTER 4

PRACTICE-, TIME-, OR TASK-BASED MODELS

This chapter describes practice-, time-, or task-based models, which are theories or models that suggest that acquisition of knowledge & skills, development of expertise, and performance improvement are a function of the nature of practice, amount of time spent on the task, and the task type. The four models are reviewed in this chapter are:

- Mastery learning model by Bloom (1968)

- Deliberate practice model of expert performance by Ericsson *et al.* (1993)

- 4-Component model of complex skill acquisition by van Merriënboer (1992)

- Recognition-primed decision-making model by Klein (1993)

4.1   BLOOM (1968) MASTERY LEARNING MODEL

Educational theorist Carroll (1963) conceptualized the 'mastery learning model' in which she challenged the traditional educational philosophy stating that 'the learner will succeed in learning a given task to the extent that he spends the time that he *needs* to learn the task' (p. 725). Carroll used certain factors like aptitude, the time required to learn the task under ideal instructions, ability to understand the instructions, and perseverance and external conditions such as the time allowed for learning and the quality of the instructions received. She speculated that a majority of learners would be successful in gaining mastery in learning using a suitable combination of these factors and systematically maximizing the time allowed for learning.

Bloom (1968, 1971; 2018) further developed the theory. Bloom (1968), in one of his experiments, noticed that, under proper learning conditions and time given to the learner, 80% of the learners demonstrated Grade A as compared to 20% of learners being able to make it to Grade A when such conditions were not made available. Guskey & Gates (1986, p. 73) summarized, 'The theory of mastery learning is based on the simple belief that *all children [learners] can learn* when provided with conditions that are appropriate for their learning.' The fundamental premise of this model is allowing time for learning. Bloom (1968, p. 7) observed that 'student must not only devote the amount of time he needs to the learning task but also that he be *allowed* enough time for the learning to take place.' He further observed that by using different instructional strategies for different learners, one could create proper conditions for all learners. Bloom (1968, p. 7) also suggested that there are 'many alternative strategies for mastery learning. Each strategy must find some way of dealing with individual differences in learners through some means of relating the instruction to the needs and characteristics of the learners.'

The key mechanism of Bloom's mastery learning model is to use an assessment that informs students (formatting assessment), identifies their difficulties (feedback), and provides remediation procedures (correctives). Guskey (2009) summarized the process as:

> With the feedback and corrective information gained from the formative assessment, each student has a detailed prescription of what more needs to be done to master the concepts or skills from the unit. This "just-intime" correction prevents minor learning difficulties from accumulating and becoming major learning problems. It also gives teachers a practical means to vary and differentiate their instruction in order to better meet students' individual learning needs. As a result, many more students learn well, master the important learning goals in each unit, and gain the necessary prerequisites for success in subsequent units. (para. 10)

The mastery learning principle has been used and demonstrated mostly in the educational setting by some researchers (Guskey & Gates 1986; Guskey 2009). However, its applications are not limited to the education industry only. Guskey (2009) cited several modern studies which showed that mastery learning could ease achieving higher-level learning goals (Arredondo & Block, 1990; Blakemore, 1992; Clark, Guskey, & Benninga, 1983; Kozlovsky, 1990; Mevarech, 1980, 1981, 1985; Mevar-ech, & Werner, 1985; Soled, 1987). In fact, Bloom (1978) emphasized the importance of problem-solving, applications of principles, analytical skills, and creativity in mastery learning approach (p. 578).

Recently, several years after its original introduction, the Bloom model resurfaced again with the name 'proficiency-based training' in several disciplines such as surgery (Brydges et al. 2012; Stefanidis et al. 2006; Scott, Ritter, Tesfay, 2008) and in military research (Salas et al. 1998). The fundamental premise of proficiency-based training is to allow trainees to practice until they demonstrate the desired standards of performance (proficiency) right during the training event, and removing the limit to time. During this period, the learners are engaged in deliberate practice through repetitive performance, are assessed rigorously, and receive informative feedback (Ericsson *et al.*, 1993). Proficiency metrics for training tasks can provide the external motivation necessary to engage the learners in the skill acquisition process beyond the usual approach of investing time on a particular task or conducting an arbitrary number of repetitions. Training programs for aircraft pilots have been recently designed and delivered using the proficiency-based training methodology by removing the restrictions on the same number of hours of practice for all trainees but to track progress by task (Stewart & Dohme 2005).

This model has been generally recognized to be useful in any level of education. The model provides strategies like differentiated learning and allows for the integration with a range of other instructional strategies. In terms of skill acquisition, proficiency/expertise development, and developing performance, the main implication of this model is that learning should be designed by considering the appropriate conditions to the performers, personalizing them based on their learning speed, and allowing enough time in the program to allow them to demonstrate their mastery to the desired standards.

However, this approach has been criticized on the grounds that time is the essence in organizational settings. Though the model is powerful, organizations usually are not at the liberty of time to wait for learners to acquire mastery at their own pace. Farr (1986, p. 65) commented, 'It assumes that the speed and mode of learning are not important. Instead, the emphasis is on the goal that each student master certain information at his own rate.' In an experiment using three pedagogical techniques to train learners in psychomotor skills, Charles & Michael (2002) noted that though participants in proficiency-based training took longer to get trained, they performed uniformly and at a higher level than the participants trained using other methods. The logistics and disparity of learners, and locational or other factors may not allow the organization to give enough time to their learners and highly individualized learning for different learners.

4.2   ERICSSON *ET AL.* (1993) DELIBERATE PRACTICE MODEL OF EXPERT PERFORMANCE

Ericsson (2009b, p. 405) negated the sole effect of experience on developing higher level expertise. He maintained that 'Experience in a domain of activity appears to be necessary to perform adequately, but extensive experience does not invariably lead people to attain superior (expert) performance.' Ericsson & Lehmann (1996) observed that the level of training and experience often have only a weak link to the objective measures of performance. In earlier research studies in the domains of music and chess, Ericsson *et al.* (1993) noticed that individuals use optimal training, deliberate professional practice, and extended domain-related activities to incrementally improve their performance. According to Ericsson (2006, p. 685), '...extensive experience of activities in a domain is necessary to reach very high levels of performance. Extensive experience in a domain does not, however, invariably lead to expert levels of achievement...further improvement depend [*sic*] on deliberate efforts to change particular aspects of performance.' Instead, individual needs 'deliberate practice' to develop expertise. Ericsson & Charness (1994) observed that individuals may be able to develop a higher level of expertise when subjected to constant engagement in similar activities, exposed to new issues in their domains, and subjected to a significant amount of deliberate practice. They maintained that deliberate practice is not just ordinary practice or any other domain-related activity like work or on-the-job training event. Ericsson (2006) brought out the differences between routine practice (that is, spending some focussed time on a task regularly) and deliberate practice. 'The core assumption of deliberate practice is that expert performance requires the opportunity to find suitable training tasks that the performer can master sequentially....typically monitored by a teacher or coach' (Ericsson 2006, p. 692). That is, 'deliberate practice' refers to highly individualized training on tasks selected by a qualified teacher to build expertise in an individual.

The deliberate practice model comprises four components: focused goals, which are determined by a teacher in order to improve a specific aspect of performance; concentration and effort; feedback from a teacher comparing actual to the desired performance; and further opportunities for practice. Feedback from a mentor is a very important factor in learning and skill acquisition. Deliberate practice activities are designed to allow the repeated experience to enable learners to observe the different critical aspects of the tasks.

Ericsson & Charness (1994) further hypothesized that the dramatic difference in the performance between experts and non-experts could be attributed to the amount of deliberate practice. Thus, the acquired performance of an individual is a direct function of the amount of time engaged in deliberate practice activities. Ericson *et al.* (1993) further found that it takes 10,000 hours or 10 years of intensive training and 'deliberate practice' to become an expert in almost anything. However, it must be noted that Ericsson's theory mostly deals with 'elite expertise' in relatively closed and repetitive domains such as sports and music. Ericsson (2004, p. S78), based on his studies in the context of medicine, opines that certain aspects of deliberate practice could be generalized to other domains by retaining the 'focus on identifying superior, reproducible behavior for representative tasks in the associated domain.' Ericsson and his colleagues coined the term 'expert-performance' to analyze the naturally occurring situations and then use them to measure performance under controlled conditions.

From the implication and application standpoints, one of the key issues with the deliberate practice model is that it is based on the repetition of the familiar activity in relatively measurable domains. Recent studies have opined that using 'deliberate practice' as the only method to achieve expertise is probably not realistic and cited the deficiencies of the approach. The deficiencies pointed out were the following: it discounted the effect of innate talent; it did not explain the performance in novel situations; it possibly ignored the effect of task complexity and task characteristics; and it exhibited a large variance of performance among individuals (Ackerman 2014; Gobet 2013; Hambrick, Altmann, *et al.* 2014; Hambrick, Oswald, *et al.* 2014; Kulasegaram, Grierson & Norman 2013; Lombardo & Deaner 2014). Most of the objections appear to stem from the issue that the deliberate practice approach is based on the repetition of familiar or routine tasks in relatively closed and repetitive domains, such as sports and music, in which the standards of measurements are defined and finite. In a domain in which problems were novel or non-repetitive in nature, the applicability of this model is not well established.

However, on the other hand, researchers have applied the concept of deliberate practice to domains such as science, weather forecasting, engineering, and military command and control (Hoffman 2007; Klein & Hoffman 1993). Sonnentag & Kleine (2000) suggest that the 'deliberate practice' mechanism proposed by Ericsson *et al.* (1993) can also be used in work settings which include activities such as extensive preparation for task accomplishment, gathering information from domain experts, and seeking feedback. They further concluded that performance in work settings is a direct function of the amount of deliberate practice. However, Sonnentag & Kleine (2000, p. 90) also caution: 'Work performance is a complex phenomenon and there are a number of factors in addition to deliberate practice that might impact an individual's work performance, for example an individual's professional experience or the effort devoted to the task. It could even be that these factors override the effects of deliberate practice.'

The major implication of this model in proficiency development is the value it places on personal efforts in terms of disciplined, deliberate practice as well as on training efforts in terms of the nature and sequence of exercises, amount of practice, and feedback or coaching mechanisms.

4.3   VAN MERRIËNBOER *ET AL.* (1992) 4-COMPONENT MODEL OF COMPLEX SKILL ACQUISITION

While most of the skill acquisition and expertise development theories and models deal with a range of contexts and skills, the major interest of researchers has been in complex cognitive skills. However, most of the models, though comprehensive, are either underserved or lack adequate guidance to training designers in designing training programs to develop complex cognitive skills. van Merriënboer, Jelsma & Paas (1992) proposed the four-component instructional design (4C/ID) model for complex learning. van Merriënboer, Clark & de Croock (2002) posit that the traditional approach to developing complex learning by training learners on oversimplified component tasks does not work. They contend that knowledge from simple tasks does not reliably apply to novel future problems. Complex learning requires a focus on total sum more than merely the constituent skills. While training learners on constituent skills and having them practice is important, the 4C/ID model maintains that learners should 'acquire the ability to use all of the skills in a coordinated and integrated fashion' while doing real-life tasks (van Merriënboer, Clark & de Croock 2002, p. 40).

In the 4C/ID model, the environment for complex learning is described in terms of four interrelated 'blueprint components— Learning tasks, Supportive information, Just-in-Time information, and Part-task practice.' Learning tasks are concert, authentic, whole-task experiences provided to learners, which primarily facilitates their induction. These are also called whole-tasks. The tasks 'confront the learners with all constituent skills that make up the whole complex skill' (van Merriënboer, Clark & de Croock 2002, p. 43). On the other hand, part-task practice refers to practice sessions on constituent tasks that aim to build rule automation (automaticity) for routine aspects of the whole complex skill. Supportive information is the bridging information based on the prior knowledge of the learners, and is provided to support the learning and performance of the nonroutine aspects of the learning tasks. Supportive information consists of mental models, cognitive strategies, and cognitive feedback. Just-in-time information refers to the prerequisite information that learners need to learn and perform routine aspects of the learning tasks. This information consists of information displays, demonstrations, and corrective feedback, and is presented to the learners as and when needed.

The premise of this model is to build the training module in such a way that it allows learners to construct rule automation for routine aspects of skills and reconstruct schemata (or mental models) for nonroutine aspects.

van Merriënboer, Clark & de Croock (2002) posit that the 4C/ID model should be used in complex skills learning when 'transfer is the overarching learning outcome' (p. 55). Several studies have shown long-term transfer of learning,; for example, computer programming (van Merriënboer, 1990a, 1990b), computer-based training studies (Schuurman, 1999; van Merriënboer & de Croock, 1992; van Merriënboer, Schuurman, de Croock, & Paas 2002), statistical analysis (Paas 1992, 1993), numerically controlled computer programming (Paas & van Merriënboer 1994), and fault management in the process industry (de Croock 1999; de Croock, van Merriënboer & Paas 1998; Jelsma 1989).

The implications of the 4C/ID model in expertise development is profound. van Gog *et al.* (2005) training techniques to reduce cognitive load during deliberate practice on routine aspects and to build proficiency. They extended it on the observation that proficient learners focus on routine rather than exceptional aspects of decision-making and problem-solving. On the other hand, van Merriënboer & Kester (2008) revolutionized the part-task technique instead of the whole-tasks approach to build expertise in pattern recognition, a characteristic of expertise. They demonstrated that pattern recognition skill, presumably developed through experience, could be taught using training when it is isolated as a part-task. In several studies conducted in the field of sports, it was observed that perceptual-cognitive skills like baseball pitch recognition could be built by isolating the complex task as part-task and designing part-task practice focussing on the particular skill (Fadde 2010; Farrow & Abernethy 2002; Ward *et al.* 2008; Williams, Ward & Chapman 2003).

4.4 KLEIN (1993) RECOGNITION-PRIMED DECISION (RPD) MODEL FOR RAPID DECISION-MAKING EXPERTISE

Professionals traditionally used analytical techniques to make decisions. However, in fast-paced and risky situations, analytical processing does not work (Simpson 2001). Such professions include military combat, aviation, law enforcement, firefighting, and surgery etc. Developing the expertise of professionals in these areas to enable them to make fast and naturalistic decisions is of paramount importance. Klein (1998) developed the *Recognition-Primed Decision* (RPD) method after interviewing fireground commanders and a number of other high-risk professionals who encounter several dynamic situations in their professional lives to understand how they make decisions and evaluate options. He found that experts typically make decisions by recognizing the aspects of new situations and comparing them with the earlier situations. RPD involves actions such as assessing the situation, mentally searching for familiar patterns one may have seen in the past, comparing the solutions from past, and deciding an appropriate course of action in the current situation. By implementing solutions or taking actions that had worked in the past, decision-makers are able to respond to situations faster (Kuchenbrod 2016).

Experience is the fundamental ingredient of such natural decision-making situations. Klein (1998) found that the following four qualities are important to make such decisions: situational awareness, tacit knowledge, mental simulation of future events, and self-efficacy. Situational awareness is the sense of paying acute attention to cues, events, alarms, signs, and other situational components that allow experts to think ahead. Simpson (2001) considers situational awareness as a critical component of intuitive decision-making.

Klein (1998) noted that experts use tacit knowledge in a given situation to understand or recognize the gravity of a situation.

The mental simulation was another ingredient. In a risky situation, experts run several what-if scenarios in their minds for potential effects, and this mental picture allows them to perceive relevant details and devise a possible option in a given situation.

The last component noted by Klien is regarding the confidence of the decision based on knowledge representation they have in their mind. This leads to intuitive decisions. Klein (1998) noted that intuition grows out of the experience when the environment provides sufficient cues and individual learns from the cues.

The Klein (1993) model reiterates the need for the appropriate execution of cognitive task analysis (CTA) in modeling experts and in capturing how they operate or make decisions. The other implication of this model is that it is imperative to train professionals on strategies like pattern recognition, in addition to analytical skills. This model has generated several pieces of evidence that expertise in decision-making can be accelerated.

♣ ♣ ♣ ♣

CHAPTER 5

FACTOR-BASED MODELS

This chapter describes the factor-based models which are based on theories or models that suggest the interplay of several factors that influence the acquisition of knowledge & skills, development of expertise, and performance improvement. The following key models are reviewed in this chapter:

- Model of intelligence as expertise development by Sternberg (1999)

- Skilled performance theory by Langan-Fox *et al.* (2002)

5.1   STERNBERG (1999) MODEL OF INTELLIGENCE AS EXPERTISE DEVELOPMENT

The model of deliberate practice presented by Ericsson *et al.* (1993) as the sole means of attaining expert performance had its own share of criticisms. For example, Sternberg (1998a) disagrees that deliberate practice is the exclusive aspect of the acquisition of expertise. He asserted that developing proficiency or expertise involves five key elements: metacognitive skills, learning skills, thinking skills, knowledge, and motivation. Sternberg (1999) reasoned that while deliberate practice is required to work toward expertise, motivation is the key to practice. Sternberg (1999, p. 364) maintains that 'the novice works toward expertise through deliberate practice. But this practice requires interaction of all five of the key elements. At the center, driving the elements is motivation.' Motivation drives the emergence of metacognition, which, in turn, activates learning as well as thinking skills. Both learning and thinking skills then enhance declarative and procedural knowledge, enabling the level of expertise to increase.

The contributing elements of Sternberg's model are described below:

1. **Metacognitive skills:** Metacognitive skills refer to the ability of individuals to self-regulate their learning. Sternberg (1985, 1986) has indicated seven metacognitive skills which are particularly important: problem recognition, problem definition, problem representation, strategy formulation, resource allocation, monitoring of problem-solving, and evaluation of problem-solving (Sternberg 1998a). The important postulation of this model is that all of these skills are modifiable and trainable (Sternberg & Spear-Swerling 1996; Sternberg & others 1985; Sternberg 1988, 1998b).

2. **Learning skills:** Learning skills help individuals attain knowledge. Sternberg (1998a) specifies two kinds of learning skills: explicit and implicit. While explicit learning refers to the learning that occurs when individuals make efforts deliberately to learn something, implicit learning refers to the learning that individuals pick incidentally without any systematic effort. Selective coding, selective filtering, and distinguishing relevant vs. non-relevant are the learning skills which are acquirable with appropriate training.

3. **Thinking skills:** Sternberg & Spear-Swerling (1996; 2005) emphasize three kinds of thinking skills that may be attributed to developing expertise through training. These are:

   - Critical (analytical) thinking skills, which include analyzing, critiquing, judging, evaluating, comparing and contrasting, and assessing.

   - Creative thinking skills, which include creating, discovering, inventing, imagining, supposing, and hypothesizing.

   - Practical thinking skills, which include applying, using, utilizing, and practicing.

4. **Knowledge skills:** Declarative knowledge refers to 'knowing that,' such as facts, concepts, principles, and laws. Procedural knowledge, on the other hand, is 'knowing how,' such as procedures and strategies. However, Sternberg *et al.* (1995) place a lot of importance on procedural knowledge as compared to declarative knowledge.

5. **Motivation:** Motivation may be categorized into two types: achievement motivation and competence motivation. 'Achievement motivation' (McClelland *et al.* 1976) refers to the tendency of individuals to seek moderate challenges and risks to achieve accomplishments. 'Competence motivation' refers to the beliefs of individuals in their own abilities to solve the problems at hand (Bandura 1977). Sternberg postulates that experts develop competence motivation that helps them solve the difficult tasks in their domains. It may be inferred that both competence motivation and achievement motivation can be instilled through managed successes during a training course.

The interactions among five elements mentioned above are affected by, and, can, in turn, affect the context in which they operate.

That is how Sternberg (1999) postulated 'context' as the sixth element in his model. He reasoned that contextual factors could also affect the task performances of individuals. The learning environment and its ability to provide the relevant context to the learners play a significant role in gaining proficiency or expertise (Caterjon *et al.*, 2006).

The important postulation of this model is that all of these skills are both modifiable and trainable (Sternberg & Spear-Swerling 1996; Sternberg & others 1985; Sternberg 1999). This led other researchers to explore techniques and strategies to build expertise through training.

5.2   LANGAN-FOX *ET AL.* (2002) SKILLED PERFORMANCE THEORY

In Chapter 3, three-stage models by Ackerman (1988), Fitts and Posner (1967), and Anderson (1982) have been reviewed. Another staged model was proposed by Schneider & Shiffrin (1977), which was not far different from the above three models. One major shortfall of all the three-stage models is that although they explain the mechanism of skill acquisition, they do not provide a way to know precisely as to where an individual is during the process of skill acquisition. As an individual starts practicing, not only his/her cognitive ability but also the task characteristics change with time and environment during the course of practice.

Notwithstanding the limitation of the Ackerman (1988), Fitts and Posner (1967), Anderson (1982), and Schneider & Shiffrin (1977) three-stage skill acquisition models, Langan-Fox *et al.* (2002) contend that one needs to consider skill acquisition as a process that encompasses changes in several other variables, internal to human external to performance, worth measuring:

Rather than examining the relationship between static measures of cognitive ability and performance at different stages of practice, it would be far more interesting to tap into changing relationships between emotions or metacognitions and performance; for example, what different patterns emerge, and how can this be generalized to other tasks or situations? (Langan-Fox *et al.* 2002, p. 107).

Skill acquisition theories have provided closely resembling accounts of the final stage of performance, expressed as *autonomous performance* (Fitts & Posner 1967) or *procedural performance* (Anderson 1981, 1982). This stage is characterized by speed and accuracy as also indicated in the Ackerman (1988) model. Langan-Fox *et al.* (2002) termed this state as 'skilled performance.'

Langan-Fox *et al.* (2002, p. 107) proposed an integrated process model for 'skilled performance' based on learning in general and during skill acquisition and considered two sets of factors that influenced skilled performance.

Internal influences: Langan-Fox *et al.* (2002) believed that factors such as changing levels of consciousness, automaticiy, motivation, emotions, metacognition, and memory influenced skill acquisition.

a) Changing levels of consciousness is based on Rasmussen's (1983) observations that a performer may make a conscious/voluntary choice among the skill-based, rule-based, and knowledge-based behaviors, depending on the task characteristics.

b) The amount of practice determines how soon an individual will achieve automaticity or automatic processing (Shiffrin & Schneider 1977). However, task inconsistency demands a higher level of practice (Ackerman 1988).

c) Langan-Fox *et al.* (2002) believe that motivation in the form of setting up difficult goals (Ackerman 1988) influences the acquisition of complex skills.

d) Self-regulatory processes and metacognition have been found to influence skill acquisition in terms of how people perceive their skills with respect to the task difficulty. This observation indicates that performers are always able to monitor their cognitive state and change their cognitive strategies during skill acquisition.

e) Langan-Fox *et al.* (2002) also postulated that emotions have a great impact on skilled performance, a factor previously missing from the skill acquisition theories but 'most central and pervasive aspects of human experience' (p. 109). Emotions may have a positive or a disruptive effect on performance depending on the affective state of the performer. It is also believed that emotions and motivations are related, though not included in previous models.

f) Memory is another component that has been found to influence skill acquisition profoundly, as explained by Anderson (1981, 1982) in the ACT-R theory. Langan-Fox *et al.* (2002) predicted that in stage 1 of skill acquisition, where demand for attentional resources is more, working memory is the most predictive of the task performance. However, in stages 2 and 3 of skill acquisition, associative memory and procedural memory, respectively, are likely to more predictive of the task performance (ibid., 111).

g) Retention factors also have been found to have a strong influence on skill acquisition because retention leads to 'improved capacity to draw an association between new and established information' (ibid., 111). All of these factors, to a varying degree, influence how individuals progress through skill acquisition and demonstrate skilled performance.

External influences: Langan-Fox *et al.* (2002) observed that external factors such as interruptions, goals, practice format, and task characteristics influenced how someone acquires skilled performance.

a) Langan-Fox *et al.* (2002) observed that the nature of the practice format (spaced or continuous) determined the time in which the state of automaticity was achieved; the interval between the practice sessions was found to affect the performance markedly.

b) Workplace interruptions and expectations to handle two or more incompatible tasks at the same time were found to adversely affect stages 1 and 2 of skill acquisition.

c) The nature of the task (novel, complex, or familiar) had an implication similar to that using cognitive resources, and, thus, impair performance (Ackerman 1988).

The major implication of this model on expertise development is that it allows to view skill acquisition as a process rather than stages. As a result, it may be concluded that several factors influence how soon an individual can reach the skilled performance stage. Based on this predictive value, it may be inferred that this model could also suggest that more than the task characteristics and practice amount, several other internal and external factors must be considered to accelerate skill acquisition.

♣ ♣ ♣ ♣

CHAPTER 6

EXPERT MODELING-BASED MODELS

This chapter describes modeling-based models, which are theories or models that suggest modeling of experts through elicitation or guidance for the acquisition of knowledge & skills, development of expertise, and performance improvement. The following key models are discussed here:

- Cognitive apprenticeship model by Collins et al. (1989)

- The trajectory of expertise by Lajoie (2003)

- Cognitive task analysis (CTA) approach to model expertise

- Zone of proximal development by Vygotsky (1978)—*already explained in Chapter 3*

6.1   COLLINS (1989) COGNITIVE APPRENTICESHIP MODEL

Collins *et al.* (1989) developed a method of direct modeling of experts in performing complex cognitive tasks. They called the approach 'cognitive apprenticeship,' which may be defined as 'learning through guided experience on cognitive and metacognitive, rather than physical, skills and processes' (Collins, Brown & Newman 1989, p. 456). Fundamentally, the cognitive apprenticeship model challenged the traditional apprenticeship model in delivering cognitive skills in situated settings. While learners build their skills under a mentor by direct observation in traditional apprenticeship, the process becomes challenging when cognitive skills, such as decision-making, are involved because most of the thinking is internal to the performer. The main premise of this model is to make thinking visible through actions or other means. As stated, 'applying apprenticeship methods to largely cognitive skills requires the externalization of processes that are usually carried out internally' (Collins, Brown & Newman 1987, p. 4). From the process standpoint, the methodology involves breaking the larger or complex tasks into smaller tasks which are within 'zone of proximal development' (Vygotsky 1978) or within reach of learners, and then supporting and providing guidance to them in situated settings on authentic and representative tasks, rather than classroom-type tasks.

The cognitive apprenticeship model proposed by Collins, Brown & Newman (1989) consists of six key components: modeling, coaching, scaffolding (and fading), articulation, reflection, and exploration.

- *Modeling:* Modeling cognitive task requires an expert to externalize the process or actions as much as possible. The process of externalization includes revealing the train of thoughts, factual and conceptual knowledge being used at a given moment, strategic knowledge like heuristics, and control strategies for making decisions. This stage also emphasizes the need to reveal the knowledge structure of experts.

- *Coaching:* Coaching refers to observing learners and offering them prompts, providing feedback, asking questions, and redirecting them to develop performance.

- *Scaffolding:* Scaffolding refers to the support experts provide to the learners while doing tasks. These include suggestions, the actual execution of part of the tasks, or cues. In both coaching and scaffolding, the experts monitor the progress of the learners (Druckman & Bjork 1991).

- *Reflection:* Reflection refers to 'enabling the learner to be critical of their [sic] own performance and problem-solving processes and to compare these with those of an expert, another learner, and ultimately, an internal cognitive model of expertise' (Woolley & Jarvis 2007, p. 75). Opportunities for reflection are built during the problem-solving process.

- *Articulation:* Articulation refers to the process during which learners are encouraged to articulate, that is, explain, summarize, clarify, or ask questions.

- *Exploration:* Exploration refers to the process during which learners are made to explore new goals, formulate and test a new hypothesis, and conduct experiments. Here, the role of the expert cannot be separated from that of the learner (Druckman & Bjork 1991).

Velmahos *et al.* (2004) conducted a research study in which they tested two models, namely, traditional "see-do-teach" for one group and training designed using CTA for the other group of medical students. They found that students using the "see-do-teach" model demonstrated serious, life-threatening errors and omissions while students trained with training designed using CTA did not make any such mistake. Supporting the concept of model-based feedback as a way to develop the expertise of novices, Fadde (2009) cited an example:

> In the study reported by Ifenthaler [2009], model-based feedback was automatically provided by a computerized system that generated a mental map based on learners' comprehension of a text passage on climate change. An expert's mental map based on the same passage was also generated and learners were presented with both maps (model-based feedback). The experiment compares the effectiveness of a full comparison of expert and learner-generated mental maps with a representation of both maps that depict only the differences between the expert's and the learners' mental maps. The underlying theory is that learners will implicitly come to think more like experts after being presented with mental-model feedback, especially in formats that highlight differences.

Various studies attempted cognitive apprenticeship in situated settings and reported positive results. Recently, more and more technology-based cognitive apprenticeship models have been demonstrated to provide expert guidance through the computer or web. Dennen & Burner (2008, p. 436), based on their review, concluded:

> Empirical studies have confirmed much of what theories have suggested that: (1) the cognitive apprenticeship model is an accurate description of how learning occurs naturally as part of everyday life and social interactions and (2) the instructional strategies that have been extracted from these observations of everyday life can be designed into more formal learning contexts with a positive effect.

From the expertise development perspective, the cognitive apprenticeship model is probably the most widely used model, in some shape and form, as a workplace training alternative. A more recent development in S-OJT is a form of the cognitive apprenticeship model, though generally used for jobs that are more predictable, routine, and can be documented (Jacobs 2001, 2003). The cognitive apprenticeship model has demonstrated evidence of shortening the time to proficiency and accelerating the development of expertise.

6.2   LAJOIE (2003) TRAJECTORY OF EXPERTISE

The key question that several researchers have been asking is 'how to develop expertise in a novice?' Lajoie (2003) suggests that the key to developing expertise is first understanding what experts know and then mapping that to create a trajectory that can help novices develop similar competencies (p. 21). She offers three different approaches to develop expertise: (a) explicitly define transitions in expertise examples and engage experts to guide and scaffold novices; (b) include dynamic forms of assessment that can create learning opportunities, such as Sherlock computer-based tutoring system that provides assessment in the context of tasks; and (c) provide explicit exemplars or models of expertise to novices, which may include responses or examples from experts (Lajoie 2003, p. 22).

Lajoie suggests that 'models of expertise that include different trajectories to competence can be used to design instruction and assessment for both in- and out-of-school contexts' (p. 21). This model of expertise could be developed using techniques like CTA to understand how experts operate, followed by mapping it in the form of a trajectory to understand possible transition points that require instruction. The model can then be used to formulate instructional design instruments to help transition novices toward expertise. 'Models of expertise can assist us in determining what to monitor, how to assess, and where to scaffold learners so that they eventually become independently proficient in their chosen fields' (p. 22).

Lajoie (2003, p. 21) concluded that 'A major finding in our work is that there is more than one trajectory to competence.' This is an important postulation that necessitates applying techniques like CTA to identify trajectories.

6.3 COGNITIVE TASK ANALYSIS (CTA) APPROACH TO MODEL EXPERTISE

Task analysis was an approach that was traditionally used to break the components of a job and to train novices. Clark & Estes (1996, p. 403) maintain that '[o]nce a job and its component tasks had been analyzed and recorded, inexperienced workers could be more quickly trained to perform necessary jobs.' Researchers have always been interested in understanding the mechanisms of how individuals acquire declarative and procedural knowledge. In simple terms, Clark & Estes (1996, p. 405) believed that 'human beings are capable of acquiring declarative knowledge, production knowledge, or both about any task. Declarative knowledge is information about "why or what". Procedural or production knowledge is information about "how and when".' Research confirms that experts develop automaticity, which is 'implicit or unconscious' (Clark & Estes 1996, p. 406) as a result of the repeated use of declarative knowledge, which, in turn, develops into production knowledge (Anderson, 1983). Most of the expert knowledge is automated procedural knowledge. However, capturing this knowledge is the real challenge. This challenge led researchers to develop approaches like CTA as a 'set of methods and techniques that specify the cognitive structures and processes associated with task performance' (Clark & Estes 1996, p. 407) to capture how experts operate, think, behave, and process a task. The fundamental premise of this approach is that knowledge takes different forms, which enable different performances at different stages of expertise. The key to expertise development is to first elicit this knowledge structure or mental models and then teach the technique to novices to help them develop expertise.

As an example, the CTA technique was used by the Federal Aviation Administration (FAA) to develop training programs for air traffic controllers (Seamster *et al.* 1993). The FAA used three-phase analysis — pair problem-solving, mental modeling and task decomposition, and strategy analysis. During this analysis, a cognitive hierarchy map of the mental model used by expert air traffic controllers was created. The information was used to organize the chunks of related information, followed by a sequence of simulation-based practice (Seamster *et al.* 1993). Leveraging the mental model and strategies used by experts to solve novel problems, training design could design a training program for less-experienced learners. It served as an incomparable way to develop the proficiency and expertise of novices.

Merkelbach & Schraagen (1994) presented three frameworks, namely, task modeling, knowledge modeling, and cognitive modeling, to develop an integrated view of how experts perform tasks, how they think, and how they approach problems. They explain these frameworks as follows:

> The first view, the task modeling view takes the task as its focus. It gives an answer to the question: "what goals have to be accomplished by performing the task?". the second view, knowledge modeling, takes the task requirements, in terms of the required knowledge and the required strategies, as its focus. The important questions in knowledge modeling are: "which performance norms are given" and "what strategies and knowledge should be used in order to accomplish the goals?". the third view, cognitive modeling, takes the performer of the task as its focus. The central question in

cognitive modeling is: "what performance is displayed in practice in order to accomplish the goals?" It gives answers to the question: "what strategies and knowledge are actually used in order to accomplish the goals?" Merkelbach & Schraagen (1994, p. 10).

These frameworks are based on the assumption that human expertise is based on knowledge structures. The model of expertise constitutes four different types of knowledge: domain knowledge, inference knowledge, task knowledge, and strategic knowledge.

Clark *et al.* (2008) assert that experts' cognitive processes can be accurately adapted in training materials, which allows for more effective expertise development. For example, Schaafstal *et al.* (2000) noticed that participants of the radar troubleshooting course developed using CTA solved more than twice the number of problems and in lesser time than those of the radar troubleshooting course developed using the traditional method. Merrill (2002) found that, in a course designed using CTA meant to teach spreadsheets, the participants scored much higher and took much lesser time in completing the tasks. Several other research studies have reiterated the benefits of the CTA approach toward effective instruction, higher performance, and reduced time to performance (Champion *et al.* 2014; Clark, Pugh, *et al.* 2008; Feldon *et al.* 2010; Katz, Hall & Lesgold 1997). In the recent past, it has been observed that CTA has become the standard toolset in research and in cognitive engineering practice. Roth & O'Hara (2014, p. 320) maintain that 'CTAs are now routinely used to understand the cognitive and collaborative demands that contribute to performance problems, the basis of expertise, as well as the opportunities to improve performance through new forms of training, user interfaces, or decision aids.' Clark, Feldon, *et al.* (2008) assert that CTA and 4C/ID models leverage each other's strengths toward the development of expertise.

Over the years, over 100 different methods have been introduced to perform CTA (Clark, Feldon, *et al.* 2008). The important ones include hierarchical task analysis (HTA); goals, operators, methods, selections (GOMS); knowledge analysis and documentation system (KADS); precursor or reason for action, action, result, interpretation of result (PARI); and the integrated task analysis model (ITAM). Hoffman & Lintern (2006) suggested three different techniques to perform CTA, which included the critical decision method (CDM), work domain analysis (WDA), and concept mapping.

More recently, Squires *et al.* (2011) used a combination of CTA methods to develop competency maps or taxonomies of expertise in system engineering jobs and to design the 'experience accelerator' simulator. Cooke (1994) identified three broad families of techniques: (1) observation and interviews, (2) process tracing, and (3) conceptual techniques.

Particularly for expertise development, Lajoie (2003, p. 24) maintains that 'Cognitive methodologies, such as CTA, can lead to the identification of expertise trajectories as well as possible transition points where instruction is needed.... The transition from student to expert professional can be accelerated when a trajectory for change is plotted and made visible to learners.' This model has provided several evidence of accelerating proficiency of learners.

6.4 VYGOTSKY (1978) ZONE OF PROXIMAL DEVELOPMENT

Please refer to Chapter 3 for the details on this model.

♣ ♣ ♣ ♣

CHAPTER 7

COGNITION-BASED
MODELS

This chapter describes the cognition-based models, which are theories or models focusing on the mechanisms of cognitive learning for the acquisition of knowledge & skills, development of expertise, and performance improvement. Two key models, apart from the ACT-R theory, are discussed in this chapter:

- ACT-R theory of cognition by Anderson (1983) – *already discussed in Chapter 3*

- Cognitive flexibility theory by Spiro *et al.* (1990)

- Cognitive transformation theory by Klein & Baxter (2006)

7.1   SPIRO *ET AL.* (1990) COGNITIVE FLEXIBILITY THEORY

Spiro *et al.* (1988, 1987, 1992) developed the cognitive flexibility theory (CFT), which focused on the nature of learning in complex and ill-structured domains, and also dealt with the transfer of knowledge and skills beyond the initial learning situations. Spiro & Jehng (1990) consider an individual's ability to spontaneously change his/her mental model adaptively to respond to changing situational demands. Spiro *et al.* (1987) view expertise in ill-structured domains as the ability to develop flexible mental models that enable individuals to apply knowledge from multiple perspectives, depending on the situation. From an expertise-building standpoint, CFT advocates presenting content to learners from several different perspectives and themes. This allows learners to present their knowledge from multiple dimensions rather than a single conceptual dimension. CFT also postulates that developing expertise depends on how well individuals learn the process of assembling the mental model. Hoffman *et al.* (2014, p. 134) commented 'Active "assembly of knowledge" from different conceptual and case sources is more important in learning (for domains of complexity and ill-structuredness) that retrieval of knowledge structures [*sic*].'

The CFT model also emphasizes that the content should not be simplified too much and supports context-dependent knowledge. Thus, CFT advocates the constructivist view and specifies using case-based learning for knowledge construction from highly interconnected knowledge sources. This theory also warns us against the potential danger of creating misconceptions by using the traditional approach of progressively increasing complexity for the learners. However, the approach may hamper the acquisition of advanced complex knowledge. Hoffman *et al.* (2014, p. 135) advocated the 'simplifying conditions method' by Reigeluth (1999), which progressively increased complexity and incrementally introduced the interactions as well as materials to be learned.

There are only a few pieces of empirical evidence to the CFT model/theory. Most of the applications have been using cognitive hypertext in web- or computer-based training. Jonassen, Ambruso &

Olesen (1992), Jacobson & Spiro (1991), Spiro *et al.* (1987), and Jonassen *et al.* (1997) reported that learners studying material using flexibility theory were better able to transfer the principles to novel, unrelated cases.

However, Soule (2016, p. 44) indicates a limitation of the CFT theory as 'CFT does not define flexibility in a way that can be evaluated or describe what the optimum level is.' Hoffman *et al.* (2014, p. 136) second this opinion by saying that 'CFT does not say what the sweet spot is for flexibility. It just pushes for more flexibility.' However, they do propose that CFT can form a conceptual theory for accelerated proficiency when/if merged with CTT.

7.2 KLEIN & BAXTER (2009) COGNITIVE TRANSFORMATION THEORY

Klein & Baxter (2009) put forward the cognitive transformation theory (CTT), which attempts to explain the process of learning in building cognitive skills. CTT places importance on 'unlearning' the faulty or flawed mental model before learners start developing new models.

Explaining the mechanism of cognitive learning, Klein & Baxter (2009) argue that cognitive skills, unlike behavior skills, depend heavily on mental models. These models or knowledge structures or schemata are modes to organize one's knowledge. Klein & Baxter (2009) believe that cognitive skills are not about adding anything more but are about 'sense making,' that is, see and think about things differently. They assert that, in order to learn a cognitive skill, an individual needs to reorganize his/her knowledge structure, which, however, does not happen with the usual components of learning— diagnosis, learning objectives, practice, and feedback. They believe that it is difficult to diagnose the subtle aspects of cognitive skills. For instance, the learning objectives are not clear for cognitive learning because novices most likely do not possess even rudimentary mental models. They further posit that practice alone may not be sufficient for developing cognitive skills because novices may not know what they should be observing/watching for and monitoring. Further, the lag between cognitive actions and results from those actions makes it difficult for novices to get immediate feedback.

Klein and Baxter (2009) contend that in order for learners to reorganize their existing mental models, they have to first unlearn some of their existing flawed models and then substitute them with new models. Klein & Baxter (2009, p. 5) place 'unlearning' as the central mechanism to 'learning' new cognitive skills by stating:

> For cognitive learning, one of the complications facing instructors is that the flawed mental models of the students act as a barrier to learning. Students need to have better mental models in order to understand the feedback that would invalidate their existing mental models. Without a good mental model, students will have trouble making use of feedback, but without useful feedback, students will not be able to develop good mental models. That is why cognitive learning may depend on unlearning as well as learning.

Klein and Baxter (2009) opine that *unlearning* is achieved with the help of a process called *sensemaking*. They believe that unlearning is facilitated through the combination of self-reflection and feedback from the mentor.

CTT is one of the theories with several implications for developing and accelerating expertise. The theory posits that as an individual goes through the process of cognitive development, he/she starts getting fixated on the process of developing mental models (Feltovich, Spiro & Coulson 1997). Klein & Baxter (2009, p. 7) noted that 'As we move further up the learning curve or have more expertise, we have to put more and more energy into unlearning—disconfirming mental models—in order to accept better ones.' Therefore, to develop and accelerate expertise, it is imperative to figure out ways to reject flawed mental models. Hoffman *et al.* (2008) opined that 'Cognitive transformation theory describes the changes to knowledge and reasoning that proficient workers need to undergo in order to make the jump from mere proficiency to superior levels of expertise.... Cognitive transformation theory argues for the importance of unlearning— experiences that force people to lose faith in their mental models so that they can move to the next level' (p. 5-2). Hoffman *et al.* (2014, p. 136) argue that 'The implication [of CTT] is that high levels of proficiency are achieved when the practitioners has [*sic*] an ability to lose confidence in an existing mental model.' Emphasizing the importance of training in building and accelerating proficiency, Klein & Baxter (2009, p. 7) appeal that 'instruction needs to diagnose limitations in mental models, design interventions to help students appreciate the flaws in their mental models, and provide experiences to enable trainees to discover more useful and accurate mental models.'

Literature provides some evidence of how making learners recognize their flawed mental models leads to robust expertise and building new mental models (DiBello, Lehman & Missildine 2010; Schmitt 1996). Based on that, Hoffman *et al.* (2008) stated that 'core ideas of cognitive flexibility theory (what makes problems difficult for learners and the simplistic understandings that result for those learners) and cognitive transformation theory (the need for unlearning experiences) are certainly pertinent to shaping any program of accelerated learning.' (p. 5-2). Hoffman *et al.* (2014) did propose that merging CFT and CTT can result in a conceptual theory for accelerated proficiency.

Hoffman *et al.* (2014) clarified that it is not important to make novices think like experts but rather make them learn like experts if we want to

93

accelerate their paths to proficiency. They cited 'We may attempt to define the cues, patterns, and strategies used by experts, and try and develop a program to teach people to *think* like experts. A different approach to skills training is to teach people how to learn like experts' (Klein 1997, p. 37).

♣ ♣ ♣ ♣

CHAPTER 8

PHASES OF SKILL ACQUISITION: INTEGRATING VARIOUS VIEWS

One of the major topics in expertise studies is the stages novices go through while becoming experts and attaining skills beyond expertise. Every organization faces the challenge of training and developing their newly hired employees to a sufficient level of skills to perform the required jobs. During my doctoral research, I investigated the nature of phases of acquisition of skills suggested by several different models. Among them, one of the most recognized works in the specifying stages of expertise was proposed by Dreyfus and Dreyfus (1986). This model holds a high academic and professional significance due to the amount of research that has been done on it. However, the model comes in several supplements and adaptations that have been made over the years by several leading researchers. Thus, this chapter is an attempt to combine the several perspectives into one. I intend to integrate the summaries of models by various leading authors and bloggers to expand or clarify the definition and characteristics of the stages of skill development, originally proposed by Dreyfus & Dreyfus (1986). Please note that I have used the words 'stage' and 'phase' interchangeably to denote the various states through which skill development occurs.

8.1 NOVICE

This is the first stage where novice works to gain a better understanding of skills mostly through formal training. Novice continues to be unaware of the particular skills or knowledge that must be applied by the practitioner in real-world situations. The learners at this stage indicate an interest and willingness to develop the necessary skills and knowledge. During this phase, the novice learns to recognize various facts and figures pertaining to the skill as well as rules for deciding how to act on it. Novice takes these facts and figures context-free. They are trained to adhere to rules rigidly and apply them in any situation. From that perspective, they will not have much situational perception or discretionary judgment on whether to apply a given rule in a given situation (Dreyfus & Dreyfus 1980, 1981, 1984, 1986).

Some other descriptions and characteristics of this stage qualified by other researchers and practitioners are summarized as follows:

- Beginners, because they have no experience with the situations in which they are expected to perform, must depend on rules to guide their actions; following rules, however, has its limits. No rule can tell novices which tasks are most relevant in real-life situations. The novice will usually ask to be shown or told what to do (Benner 1984, pp. 13–34).

- Novices have rigid adherence to taught rules or plans. Little situational perception. No discretionary judgment (Eraut, 1994).

- Novice is literally, someone who is now – a probationary member. There has been some minimal exposure to the domain. Hoffman called another level as 'Initiate' to indicate a novice who has been through an initiation ceremony and has begun introductory instruction (Hoffman 1998).

- Novice has minimal, or 'textbook' knowledge without connecting it to practice; unlikely to be satisfactory unless closely supervised; needs close supervision or instruction; little or no conception of dealing with complexity; tends to see actions in isolation (Institute of Conservation 2003).

- Novices are beginners who lack any previous experience with a task. The novice learns basic rules for necessary actions but lacks the understanding to deviate from a prescribed performance. Therefore, novices can perform an action only by applying rules they have learned to use in a specific context (Gunderman 2009).

- At the novice level, knowledge is treated without reference to context but no recognition of relevance. Context is assessed analytically, while decision-making is rational (Lester 2010).

- Novices exhibit rules (protocol)-based performance; direct supervision is needed at all times. Unable to deal with complexity. The task is seen in isolation (Khan & Ramachandran 2012).

- Novices operate by using context-free features and rules; do not understand that rules are contextually based; context-free rules need to occasionally be violated given the context or situation presented; do not assume responsibility for the consequences. Thus the desire to create a protocol or a set of concrete rules results; follow rules (https://www.rebeccawestburns.com/my-blog-3/notes/five-stages-of-acquiring-expertise-novice-to-expert).

- A novice (or apprentice) is, by definition, new to a job. Novices know little or nothing about the work, certainly too little to be able to perform to an acceptable standard. Novices must be taught (or shown) the basics of what is to be done before they can have any chance of being productive. The learning strategy here is overwhelmingly instructional. "Show me (teach me) how to do my job," they ask (Rosenberg 2012).

- A novice is new to a work situation. There is often some but minimal exposure to the work beforehand. As a result, the individual lacks the knowledge and skills necessary to meet the requirements set to adequately perform the work (Jacobs 1997).

8.2   ADVANCED BEGINNER

As the novice attains some experience in real situations, his performance starts improving to a marginally acceptable level (DiBello, Lehman & Missildine 2010). Learners in this stage develop the comprehension of objective facts, initial concepts, and specific rules and are able to apply them within a discipline or in structured settings but may struggle to apply them to real-world situations (Noreen 1975). As novice gains more practical and concrete experience, he starts comparing the new situations with previously experienced situations but still applies the earlier learned rules. This enables him to deal with unrecognized facts and elements. At this stage, the learner learns to apply more sophisticated rules to both context-free and situation factors. These rules make it possible for advanced beginners to shape the experience so that it is possible to learn from experience, but situational perception is still limited. Learners may be comfortable solving routine well-defined problems but may be ineffective and inefficient in manipulating knowledge in unfamiliar settings or in solving ill-defined problems (Dreyfus & Dreyfus 1980, 1981, 1984, 1986).

Some other descriptions and characteristics of this stage qualified by other researchers and practitioners are summarized as follows:

- An advanced beginner is one who has coped with enough real situations to note (or to have them pointed out by a mentor) the recurrent meaningful aspects of situations. An advanced beginner needs help to set priorities since she/he operates on general guidelines and is only beginning to perceive recurrent meaningful patterns. The advanced beginner cannot reliably sort out what is most important in complex situations and will need help to prioritize (Benner 1984, pp. 13-34).

- An advanced beginner needs guidelines for action based on attributes or aspects (aspects are global characteristics of situations recognizable only after some prior experience); situational perception is still limited; all attributes and aspects are treated separately and given equal importance (Eraut 1994).

- Advanced beginners have a working knowledge of key aspects of practice; straightforward tasks likely to be completed to an acceptable standard; able to achieve some steps using own judgment, but supervision needed for the overall task; appreciates complex situations but only able to achieve a partial resolution; Sees actions as a series of steps (Institute of Conservation 2003).

- Advanced beginners have developed the ability to distinguish between more and less characteristic features of a situation, although they still tend to rely on checklists (Gunderman 2009).

- At the advanced beginner level, knowledge is treated in context but no recognition of relevance; Context is assessed analytically; While decision-making is rational (Lester 2010).

- Advanced beginners show guidelines-based performance; able to achieve partial resolution of complex tasks; task is seen as a series of steps; able to perform routine tasks under indirect supervision;

Direct supervision needed for complex tasks only (Khan & Ramachandran 2012).

- An advanced beginner has achieved considerable experience; more sophisticated rules that are situational; develops the idea that the idea of developing skill is a much larger conception; starts to recognize elements/ context-free features. The advanced beginner begins to ask the question – how? How does one (fill in the blank)?; Can set goals but cannot set them reasonably (https://www.rebeccawestburns.com/my-blog-3/notes/five-stages-of-acquiring-expertise-novice-to-expert).

8.3   COMPETENT

With experience, the learner begins to recognize more and more context-free and situational elements. At this point, the learner is able to organize the situation and then concentrate on important elements. He is able to assess the situation, set the goal, and then choose the best course of action. He may or may not apply rules. He may or may not be successful, but that constitutes an important element of future expertise (Dreyfus & Dreyfus 1980, 1981, 1984, 1986).

Some other descriptions and characteristics of this stage qualified by other researchers and practitioners are summarized as follows:

- A competent person is good at coping with crowdedness; Now sees actions at least partially in terms of longer-term goals; Conscious, deliberate planning; Standardized and routinized procedures (Eraut 1994).

- A competent person shows a good working and background knowledge of the area of practice; Fit for purpose, though may lack refinement; Able to achieve most tasks using own judgment; Copes with complex situations through deliberate analysis and planning;

Sees actions at least partly in terms of longer-term goals (Institute of Conservation 2003).

- At the competence level, in terms of knowledge (domain/topic), learners demonstrate a foundation body of knowledge. In terms of strategic processing, learners at this level use surface-level strategies and develop deep-processing strategies to acquire knowledge. Learner's individual interest increases and reduced reliance on situational interest (Alexander 2003).

- When learners achieve competence, they can think conceptually and develop strategic approaches in terms of long-term goals. Yet, in many situations, their approaches remain highly standardized and rule-based (Gunderman 2009).

- At a competent level, knowledge is treated in context, and also there is a recognition of relevance; Context is assessed analytically; While decision-making is rational (Lester 2010).

- At the competent level, the performance is not solely based on rules and guidelines but also on previous experience; Able to perform complex routine tasks; Able to deal with complexity with analysis and planning; Task is seen as one construct; Training and supervision needed for nonroutine complex tasks (Khan & Ramachandran 2012).

- A competent person has more experience; possesses a sense of importance and is able to prioritize behaviors based on levels of importance; behavior is determined by importance and not by context-free rules or merely situational rules; possess a hierarchical procedure for making decisions; requires organization and the creation of a plan; accepts responsibility for choices because they recognize they made choices; they are emotionally invested in their decision-making; problem-solving indicates competence; slow and detached reasoning (problem-solving);

Makes decisions (https://www.rebeccawestburns.com/my-blog-3/notes/five-stages-of-acquiring-expertise-novice-to-expert).

- Competent (or journeyman) workers can perform jobs and tasks to basic standards. They've had their basic training and now look for more coaching and practice to get better at what they do. "Help me do it better," is their primary request (Rosenberg 2012).

8.4 PROFICIENT

At this level, a learner is deeply involved in the task. He is capable of identifying the important part of the tasks and paying requisite attention. A proficient person sees the situations holistically in terms of various elements. As the situation changes, his deliberation, plan, and assessment may change. With changing situations, he is able to see new patterns which deviate from the normal. Decision-making is very quick and fluid because of the experience in a similar situation in the past. However, proficient learners will use maxims to guide their decision-making. Consistency in performance distinguishes this phase from the previous phase (Dreyfus & Dreyfus 1980, 1981, 1984, 1986).

Some other descriptions and characteristics of this stage qualified by other researchers and practitioners are summarized as follows:

- Proficient is someone who perceives a situation as a whole rather than in terms of parts. With a holistic understanding, decision-making is less labored since the professional has a perspective on which of the many attributes and aspects present are the important ones. The proficient performer considers fewer options and hones in on the accurate region of the problem (Benner 1984, pp. 13-34).

- A proficient person sees situations holistically rather than in terms of aspects; Sees what is most important in a situation; Perceives deviations from the normal pattern; Decision-making is less labored; Uses maxims for guidance, whose meanings vary according to the situation (Eraut 1994).

- A proficient person exhibits a depth of understanding of discipline and area of practice; fully acceptable standard achieved routinely; able to take full responsibility for own work (and that of others where applicable); deals with complex situations holistically; decision-making more confident; sees overall 'picture' and how individual actions fit within it (Institute of Conservation 2003).

- At proficiency and expertise stage, learners exhibit a broad and deep topic/domain knowledge base; use deep-processing strategies almost exclusively; high individual interest and engagement (Alexander 2003).

- Proficient learners can distinguish between typical and atypical features of a case and tailor their approach to the particular features at hand (Gunderman 2009).

- At the proficiency stage, knowledge is treated in context, and also there is a recognition of relevance; context is assessed holistically; While decision-making is still rational (Lester 2010).

- At the proficient level, performance mostly is based on experience; Able to perform on acceptable standards routinely; Able to deal with complexity analytically; Related options are also seen beyond the given task; Still needing supervision for nonroutine complex tasks; Able to train and supervise others performing routine complex tasks (Khan & Ramachandran 2012).

- A proficient person uses intuition based on enough past experience; deep situational involvement and recognition of similarity; Intuitive-based cognition coupled with detached decision-making. The proficient person recognizes intuitively but responds by more calculative decisions. Being proficient means attributing success to the calculative aspects of the success and ignoring the even more brilliant intuition that occurred first (https://www.rebeccawestburns.com/my-blog-3/notes/five-stages-of-acquiring-expertise-novice-to-expert).

8.5   EXPERT

Experts don't apply rules or use any maxims or guidelines. He rather has an intuitive grasp of situations based on his deep tacit understanding. One key aspect of this level is that an individual relies on intuition and an analytical approach is used only in new situations or unrecognized problems not earlier experienced. Experience-based deep understanding provides him with very fluid performance. At this stage, skills become automatic that even expert is not aware of it. Based on prior experience, they can even come up with the solution for new never experienced before situations (DiBello, Lehman & Missldine 2010). "Experts" adopt a contextual approach to problem-solving and understand the relative, non-absolute nature of knowledge. This ability distinguishes the "expert" from the "proficient" practitioner (D'Youville College, n.d.). Reflection comes naturally and experts solve problems almost unconsciously (Dreyfus & Dreyfus 1980, 1981, 1984, 1986).

Some other descriptions and characteristics of this stage qualified by other researchers and practitioners are summarized as follows:

- The expert professional is one who no longer relies on an analytical principle (rule, guideline, maxim) to connect an understanding of the situation to appropriate action. With an extensive background of experience, the expert has an intuitive grasp of the situation and focuses in on the accurate region of the problem without wasteful consideration of a larger range of unfruitful possibilities (Benner 1984, pp. 13-34).

- Expert no longer relies on rules, guidelines or maxims; an Intuitive grasp of situations based on deep tacit understanding; Analytic approaches used only in novel situations or when problems occur; Vision of what is possible (Eraut 1994).

- An expert is a distinguished or brilliant journeyman, highly regarded by peers, whose judgments are uncommonly accurate and reliable, whose performance shows consummate skill and economy of effort, and who can deal effectively with certain types of rare or "tough" cases. Also, an expert is one who has special skills or knowledge derived from extensive experience with subdomains (Hoffman 1998).

- Expert is someone who has an authoritative knowledge of the discipline and deep tacit understanding across an area of practice; excellence achieved with relative ease; able to take responsibility for going beyond existing standards and creating own interpretations; holistic grasp of complex situations, moves between intuitive and analytical approaches with ease; sees overall 'picture' and alternative approaches; vision of what may be possible (Institute of Conservation 2003).

- At proficiency and expertise level, learner possesses a broad and deep topic/domain knowledge base; Use deep-processing strategies almost exclusively; High individual interest and engagement (Alexander 2003).

- Expert learners do not use rules and guidelines. Their problem-solving is based on an intuitive grasp of relevant features and a conceptual understanding of underlying principles (Gunderman 2009).

- At the expert level, knowledge is treated in context, and also there is a recognition of relevance; Context is assessed holistically; While decision-making is now intuitive (Lester 2010).

- At the expert level, performance based on experience and intuition; Achieves excellent performance In complex situations moves easily between analytical and intuitive solutions; All options related to the given task are considered; Able to train and supervise others performing routine and nonroutine complex tasks (Khan& Ramachandran 2012).

- Expert functions or responds as a result of "mature and practiced understanding", loss of awareness of intuition and decision-making – operates simply because he does; knowledge becomes tacit; experts "see" but sometimes don't recognize that they "see"; experts perform without reflecting on every behavior, but experts do reflect and will consider alternatives when presented with the time and critical outcomes. When experts reflect, they engage in critical reflection of their own assumptions; actions are unconscious operating out of intuition and tacit knowledge; performance is fluid; "But when time permits and much is at stake, the detached deliberative rationality of the type described can enhance the performance of even the intuitive expert (p. 40)." (https://www.rebeccawestburns.com/my-blog-3/notes/five-stages-of-acquiring-expertise-novice-to-expert).

- Considering master and expert as one single stage, masters and experts create new knowledge. They invent new and better ways to do a job, and they can teach others how to do it. They are truly unique individuals and seek to learn in unique and personal ways, primarily through collaboration, research, and problem-solving. "I'll create my own learning," they say (Rosenberg 2012).

- An exprt is one who has the knowledge and experience to meet and often exceed the requirements of performing a particular unit of work. The individual is respected by others and highly regarded by peers because of his or her consummate skills, or expertise. The individual can use this ability to deal with routine and nonroutine cases, with an economy of effort (Jacobs 1997).

8.6 MASTERY

Dreyfus and Dreyfus (2001) included the sixth stage of "Mastery" beyond expertise in their model, stating mastery as "A very different sort of deliberation from that of a rule-using competent performer or of a deliberating expert characterizes the master." An important difference between an expert and a master is explained by Dreyfus (2001) as:

> When an expert learns, she must either create a new perspective in a situation when a learned perspective has failed, or improve the action guided by a particular intuitive perspective when the intuitive action proves inadequate. A master will not only continue to do this, but will also, in situations where she is already capable of what is considered adequate expert performance, be open to a new intuitive perspective and accompanying action that will lead to performance that exceeds conventional expertise (Dreyfus 2001, p. 44).

Some other descriptions and characteristics of this stage qualified by other researchers and practitioners are summarized as follows:

- Traditionally, a master is any journeyman or expert who is also qualified to teach those at a lower level. Traditionally, a master is one of an elite group of experts whose judgments set the regulations, standards, or ideals. Also, a master can be that expert who is regarded by the other experts as being "the" expert, or the "real" expert, especially with regard to sub-domain knowledge (Hoffman 1998).

- In contrast to experts, masters have developed recognizable personal styles of practice, like the style of a great artist or composer. They welcome novelty as an opportunity to reexamine their assumptions and explore new ways of thinking (Gunderman 2009).

- At mastery level, performance becomes a reflex in most common situations; Sets new standards of performance; Mostly deals with complex situations intuitively; Has a unique vision of what may be possible related to the given task; Able to train other experts at national or international level (Khan & Ramachandran 2012).

- The master is the one who is regarded as "the" expert among experts or the "real" expert among all employees. He or she is among the elite group whose judgments are looked upon to set the standard and ideals for others (Jacobs 1997).

8.7 RELEVANCE OF STAGES IN PROFESSIONAL JOBS

In the end, it is worth mentioning that the above definitions and characteristics are used by researchers and practitioners alike as a 'convenient' way to explain the progression of a learner and to indicate somehow where he or she is on the continuum. It must be understood that the divide of stages is artificial and may correspond to certain tasks. For example, a learner may be masterful in one task while he may be at an advanced beginner level in another task. Thus, it is not reasonable to label learners collectively or holistically with the name of the stage. A typical job requires varying levels of skillfulness to achieve the objectives of a job.

In recent doctorate research with over 85 training and workplace experts, I found that in the organizations, practitioners maintained the stand that the job-role proficiency in any job did not imply progression through different stages or levels of performance; rather it was referred to as achieving one pre-established performance level (Attri, 2018). A vast amount of literature emphasizes progression toward proficiency in the form of several stages in which each stage is described with a qualitatively different proficiency in skills (Alexander 2003b; Benner 2004; Dreyfus & Dreyfus 2005, 2009; Hoffman 1998; Jacobs 1997, 2001, 2003; Jacobs & Washington 2003). On the contrary, in the mentioned study, organizations did not appear to obtain any advantage by putting labels on their employee's development or progression in the journey toward proficiency. None of the 85 project leaders expressed any inclination toward characterizing or labeling any other in-between stage before proficiency. Main concerns appeared to be whether an individual was operating in the state of proficiency or not.

We have a similar observation by Dall'Alba and Sandberg (2006) who asserted that experience and understanding of practice were the major determinants of any professional skill development. It may not have level-like stages. This finding also raises the question of the usefulness of the meaning or representation of stages of skill/proficiency acquisition for the workplace. Though studies have either detected or forced-fit the characteristics of individual proficiency to the description of stages found in such models, it is not clear from those studies if there was any business benefit to the organization or any development benefit to the individual (Benner 2004; Beta & Lidaka 2015; Ramsburg 2010; Scobey 2006). Nevertheless, this study raises important concerns regarding whether or not proficiency tracking in the workplace should rely on staged-transition models.

♣ ♣ ♣ ♣

RELEVANT PUBLICATIONS BY THE AUTHOR

1. Attri, RK 2019a, *Speed To Proficiency in Organizations: A Research Report on Model, Practices and Strategies to Shorten Time To Proficiency*, Ebook, Speed To Proficiency Research: S2Pro©, Singapore, available at < https://www.amazon.com/gp/product/B07NYS81HQ/>.

2. Attri, RK 2019b, *Designing Training to Shorten Time to Proficiency: Online, Classroom and On-The-Job Learning Strategies from Research*, Speed To Proficiency Research: S2Pro©, Singapore, available at <https://www.amazon.com/gp/product/9811406324>.

3. Attri, RK 2018, *Accelerating Complex Problem-Solving Skills: Problem-Centered Training Design Methods*, Ebook, Speed To Proficiency Research: S2Pro©, Singapore, available at < https://www.amazon.com/gp/product/B07P6CXXJL/>.

4. Attri, RK 2018, *Accelerate your leadership development in training domain: Proven success strategies for new training & learning managers*, Speed To Proficiency Research: S2Pro©, Singapore, available at <https://www.amazon.com/ /dp/9811400660/>.

5. Attri, RK 2014, "Rethinking professional skill development in competitive corporate world: accelerating time-to-expertise of employees at workplace," in J Latzo (ed.), *Proceedings of Conference on Education and Human Development in Asia*, Hiroshima, 2-4 March, PRESDA Foundation, Kitanagova, pp. 1–11, http://dx.doi.org/10.13140/RG.2.1.5125.7043.

6. Attri, RK 2018, "Modelling accelerated proficiency in organisations: practices and strategies to shorten," PhD thesis, Southern Cross University, Lismore, Australia.

7. Attri, RK & Wu, WS 2018, "Model of accelerated proficiency in the workplace: six core concepts to shorten time-to-proficiency of employees," *Asia Pacific Journal of Advanced Business and Social Studies*, vol. 4, no. 1, http://dx.doi.org/10.25275/apjabssv4i1bus1.

8. Attri, RK & Wu, WS 2017, "Model of accelerated proficiency in the workplace: six core concepts to shorten time-to-proficiency of employees," *First Australia and New Zealand Conference on Advanced Research (ANZCAR)*, Melbourne, Asia Pacific Institute of Advanced Research, Melbourne, 17-18 June, pp. 1-10, viewed 24 July 2017, <http://apiar.org.au/wp-content/uploads/2017/07/1_ANZCAR_2017_BRR713_Bus-1-10.pdf>.

9. Attri, R. K. & Wu, W. S. (2015). E-Learning Strategies to Accelerate Time-to-Proficiency in Acquiring Complex Skills: Preliminary Findings. Paper presented at *E-learning Forum Asia Conference*, Jun 2015. Singapore: SIM University, available at <https://www.researchgate.net/publication/282647943>.

10. Attri, RK & Wu, W 2015, 'Conceptual model of workplace training and learning strategies to shorten time-to-proficiency in complex skills: preliminary findings,' paper presented to the *9th International Conference on Researching in Work and Learning (RWL)*, Singapore, 9-11 December, viewed 24 June 2017, <https://www.researchgate.net/publication/286623558>.

11. Attri, RK & Wu, WS 2016a, "Classroom-based instructional strategies to accelerate proficiency of employees in complex job skills," paper presented to the Asian American Conference for Education, Singapore, 15-16 January, viewed 24 June 2017, <https://www.researchgate.net/publication/303803099>.

12. Attri, RK & Wu, WS 2016b, "E-learning strategies at workplace that support speed to proficiency in complex skills," in M Rozhan

and N Zainuddin (eds.), *Proceedings of the 11th International Conference on E-Learning: ICEl2016*, Kuala Lumpur, 2–3 June, Academic Conference and Publishing, Reading, pp. 176–184, viewed 24 June 2017,
<https://www.researchgate.net/publication/303802961>.

♣ ♣ ♣ ♣

REFERENCES

1. Ackerman, PL 1988, "Determinants of individual differences during skill acquisition: cognitive abilities and information processing," *Journal of Experimental Psychology General*, vol. 117, no. 3, pp. 288–318, http://dx.doi.org/10.1037/0096-3445.117.3.288.

2. _____ 1992, "Predicting individual differences in complex skill acquisition: dynamics of ability determinants," *Journal of Applied Psychology*, vol. 77, no. 5, pp. 598–614, http://dx.doi.org/10.1037/0021-9010.77.5.598.

3. _____ 2014, "Nonsense, common sense, and science of expert performance: talent and individual differences," *Intelligence*, vol. 45, no. 4, pp. 6–17, http://dx.doi.org/10.1016/j.intell.2013.04.009.

4. Alexander, PA 1997, "Mapping the multidimensional nature of domain learning: the interplay of cognitive, motivational, and strategic forces" M Maehr and P Pintrich (eds.), *Advances in Motivation and Achievement*, pp. 213–250.

5. _____ 2003a, "The development of expertise: the journey from acclimation to proficiency," *Educational Researcher*, vol. 32, no. 8, pp. 10–14, http://dx.doi.org/10.3102/0013189X032008010.

6. _____ 2003b, "Can we get there from here?," *Educational Researcher*, vol. 32, no. 8, pp. 3–4, http://dx.doi.org/10.3102/0013189X032008003.

7. Anderson, J 1990, *The adaptive character of thought*, Lawrence Erlbaum, Hillsdale.

8. Anderson, J & Lebiere, C 1998, *The atomic components of thought*, Lawrence Erlbaum, NJ.

9. Anderson, JR 1981, *Acquisition of Cognitive Skill*, Report No. 81-1, Carnegie-Mellon University, viewed 24 June 2017, <http://www.dtic.mil/cgi-bin/GetTRDoc?AD=ADA103283>.

10. _____ 1982, "Acquisition of cognitive skill," *Psychological Review*, vol. 89, no. 4, pp. 369–406, http://dx.doi.org/10.1037/0033-295X.89.4.369.

11. _____ 2000, *Learning and memory*, John Wiley, New York.

12. Avolio, BJ, Waldman, DA & McDaniel, MA 1990, "Age and work performance in nonmanagerial jobs: the effects of experience and occupational type,"

REFERENCES

Academy of Management Journal, vol. 33, no. 2, pp. 407–422, http://dx.doi.org/10.2307/256331.

13. Baker, R 2006, "The development of expertise: the journey from acclimation to proficiency, a critical review," *Journal of Comprehensive Research*, vol. 4, p. 48, viewed 24 June 2017, <http://www.aabri.com/papers/JCR06-2.pdf>.

14. Bandura, A 1977, "Self-efficacy: toward a unifying theory of behavioral change.," *Psychological Review*, vol. 84, no. 2, p. 191, viewed 24 October 2018, <http://citeseerx.ist.psu.edu/viewdoc/download?doi=10.1.1.315.4567&rep=rep1&type=pdf>.

15. Bedi, A 2003, "Student profiling: the dreyfus model revisited," *Education for Primary Care*, vol. 14, no. 3, pp. 360–363, viewed 24 June 2017, <https://www.researchgate.net/publication/293527610>.

16. Benner, P 1984, *From novice to expert: excellence and power in clinical nursing practice*, Addison-Wesley, Palo Alto, http://dx.doi.org/10.1097/00000446-198412000-00025.

17. _____ 2001, *From novice to expert: excellence and power in clinical nursing practice*, Commemorative edn, Prentice Hall, London, http://dx.doi.org/10.1097/00000446-198412000-00025.

18. _____ 2004, "Using the dreyfus model of skill acquisition to describe and interpret skill acquisition and clinical judgment in nursing practice and education," *Bulletin of Science, Technology and Society*, vol. 24, no. 3, pp. 188–199, http://dx.doi.org/10.1177/0270467604265061.

19. Billett, S 2000, "Guided learning at work," *Journal of Workplace Learning*, vol. 12, no. 7, pp. 272–285, viewed 24 October 2018, <http://www.oce.uqam.ca/wp-content/uploads/2014/12/1403_billett-_2000.pdf>.

20. Bjork, RA 2009, Structuring the conditions of training to achieve elite performance: reflections on elite training programs and related themes in chapters 10-13, in K Ericsson (ed.), *Development of professional expertise: Toward measurement of expert performance and design of optimal learning environments*, Cambridge University Press, New York, pp. 312–329, http://dx.doi.org/10.1017/cbo9780511609817.017.

21. Bloom, BS 1968, "Learning for mastery," *UCLA Evaluation Comment*, vol. 1, no. 2, pp. 1–12, viewed 24 June 2017, <http://ruby.fgcu.edu/courses/ikohn/summer/pdffiles/learnmastery2.pdf>.

22. _____ 1971, *Mastery learning: Theory and practice*, J Block (ed.), Holt, Rinehart and Winston, New York.

23. Bransford, JD, Brown, AL, Cocking, RR, Donovan, MS & Pellegrino, JW (eds) 2004, *How people learn brain, mind, experience, and school*, Expanded edn, National Academy Press, Washington, D.C., http://dx.doi.org/10.17226/6160.

24. Brydges, R, Nair, P, Ma, I, Shanks, D & Hatala, R 2012, "Directed self-regulated learning versus instructor-regulated learning in simulation

training," *Medical Education*, vol. 46, no. 7, pp. 648–656, http://dx.doi.org/10.1111/j.1365-2923.2012.04268.x.

25. Campbell, JP 1990, Modeling the performance prediction problem in industrial and organizational psychology, in M Dunnette & L Hough (eds.), *Handbook of industrial and organizational psychology, Vol. 1*, Consulting Psychologists Press, Palo Alto, pp. 687–732.

26. Carlson, RA, Khoo, BH, Yaure, RG & Schneider, W 1990, "Acquisition of a problem-solving skill: levels of organization and use of working memory.," *Journal of Experimental Psychology: General*, vol. 119, no. 2, pp. 193–214.

27. Carroll, JB 1963, "A model of school learning," *Teachers College Record*, vol. 64, pp. 723–733.

28. Carroll, WM 1994, "Using worked examples as an instructional support in the algebra classroom.," *Journal of Educational Psychology*, vol. 86, no. 3, pp. 360–367.

29. Champion, M, Jariwala, S, Ward, P & Cooke, NJ 2014, "Using cognitive task analysis to investigate the contribution of informal education to developing cyber security expertise," *Proceedings of the Human Factors and Ergonomics Society 58th Annual Meeting*, Chicago, 27-31 October, SAGE, Thousand Oaks, pp. 310–314, http://dx.doi.org/10.1177/1541931214581064.

30. Chapman, A n.d., "Conscious competence learning model matrix – unconscious incompetence to unconscious competence,", viewed 24 October 2018, <https://www.businessballs.com/self-awareness/conscious-competence-learning-model-63/>.

31. Charles, E & Michael, J 2002, "Retraining climate as a predictor of retraining success and as a moderator of the relationship between cross-job retraining time estimates and time to proficiency in the new job," *Group \& Organization Management*, vol. 27, no. 2, pp. 294–316, http://dx.doi.org/10.1177/10501102027002007.

32. Chase, WG & Simon, HA 1973, "Perception in chess," *Cognitive Psychology*, vol. 4, no. 1, pp. 55–81, http://dx.doi.org/10.1016/0010-0285(73)90004-2.

33. Cheetham, G & Chivers, GE 2005, *Professions, competence and informal learning*, Edward Elgar Publishing.

34. Chi, MT 2006, Two approaches to the study of experts' characteristics, in R Hoffman, N Charness & P Feltovich (eds.), *The Cambridge Handbook of Expertise and Expert Performance*, Cambridge University Press, New York, pp. 21–30, http://dx.doi.org/10.1017/CBO9780511816796.002.

35. Chi, MT, Glaser, R & Farr, M (eds) 1988, *The nature of expertise*, Lawrence Erlbaum, Hillsdale, http://dx.doi.org/10.4324/9781315799681.

36. Chi, MT, Glaser, R & Rees, E 1982, Expertise in problem solving, in R Sternberg (ed.), *Advances in psychology of human intelligence, Vol.1*, Erlbaum, Hillsdale, pp. 7–75, viewed 24 June 2017, <http://www.dtic.mil/cgi-

REFERENCES

bin/GetTRDoc?AD=ADA100138>.

37. Clark, RE 2006, How much and what type of guidance is optimal for learning from instruction?, in S Tobias & T Duffy (eds.), *Constructivist Theory Applied to Instruction: Success or Failure*, Routledge, New York, pp. 158–183, viewed 24 June 2017, <http://www.anitacrawley.net/Articles/2009How%20much%20and%20what%20type%20of%20guidance%20is%20optimal%20for%20learning%20from%20instruction.pdf>.

38. _____ 2010, "Recent neuroscience and cognitive research findings on cyber learning," paper presented to the AECT Annual Convention, Anaheim, viewed 24 June 2017, <http://www.cogtech.usc.edu/publications/clark_aect_oct_28_2010.pdf>.

39. Clark, RE & Estes, F 1996, "Cognitive task analysis for training," *International Journal of Education Research*, vol. 25, no. 5, pp. 403–417, http://dx.doi.org/10.1016/S0883-0355(97)81235-9.

40. Clark, RE, Feldon, DF, van Merriënboer, JJG, Yates, KA & Early, S 2008, Cognitive task analysis, in J Spector, M Merrill, J van Merriënboer & M Driscoll (eds.), *Handbook of research on educational communications and technology*, Lawrence Erlbaum, Mahwah, pp. 577–593, viewed 24 June 2017, <http://www.learnlab.org/research/wiki/images/0/0b/Clarketal2007-CTAchapter.pdf>.

41. Clark, RE, Pugh, CM, Yates, K & Sullivan, M 2008, *The Use of Cognitive Task Analysis and Simulators for after Action Review of Medical Events in Iraq*, Center for Cognitive Technology, University of California, Los Angeles, viewed 24 June 2017, <http://www.dtic.mil/cgi-bin/GetTRDoc?AD=ADA466686>.

42. Clavarelli, A, Platte, WL & Powers, JJ 2009, "Teaching and assessing complex skills in simulation with application to rifle marksmanship training," *Interservice Industry Training, Simulation and Education Conference (I/ITSEC)*, Orlando, National Training and Simulation Association (NTSA), Arlington, viewed 24 June 2017, <http://www.dtic.mil/cgi-bin/GetTRDoc?AD=ADA535072>.

43. Collins, A, Brown, JS & Newman, SE 1987, *Cognitive Apprenticeship: Teaching the Craft of Reading, Writing and Mathematics*, Technical Report No. 403, Center for the study of reading. University of Illinois, Urbana, Champaign, viewed 24 June 2017, <http://files.eric.ed.gov/fulltext/ED284181.pdf>.

44. Collins, A, Brown, JS & Newman, SE 1989, Cognitive apprenticeship: teaching the craft of reading, writing and mathematics, in L Resnick (ed.), *Knowing, learning, and instruction: Essays in honor of Robert Glaser*, Lawrence Erlbaum, Hillsdale, pp. 453–494, viewed 24 June 2017, <http://www.dtic.mil/cgi-bin/GetTRDoc?AD=ADA178530>.

45. Collins, H 2011, "Three dimensions of expertise," *Phenomenology and the Cognitive Sciences*, vol. 12, no. 2, pp. 253–273, http://dx.doi.org/10.1007/s11097-011-9203-5.

46. Collins, H, Evans, R, Ribeiro, R & Hall, M 2006, "Experiments with interactional expertise," *Studies in History and Philosophy of Science Part A*, vol. 37, no. 4, pp. 656–674, http://dx.doi.org/10.1016/j.shpsa.2006.09.005.

47. Cooke, NJ 1994, "Varieties of knowledge elicitation techniques," *International Journal of Human-Computer Studies*, vol. 41, no. 6, pp. 801–849, viewed 24 October 2018, <http://www.academia.edu/download/35348307/knowledgeelicitation.pdf>.

48. Cooper, G & Sweller, J 1987, "Effects of schema acquisition and rule automation on mathematical problem-solving transfer.," *Journal of Educational Psychology*, vol. 79, no. 4, pp. 347–362.

49. Cornford, I & Athanasou, J 1995, "Developing expertise through training," *Industrial and Commercial Training*, vol. 27, no. 2, pp. 10–18, http://dx.doi.org/10.1108/00197859510082861.

50. Day, J 2002, "What is an expert?," *Radiography*, vol. 8, no. 2, pp. 63–70, viewed <http://www.blumehaiti.org/uploads/2/8/3/8/2838360/fallcelloday.pdf>.

51. Dennen, VP & Burner, KJ 2008, The cognitive apprenticeship model in educational practice, in J Spector, M Merrill, J van Merriënboer & M Driscoll (eds.), *Handbook of research on educational communications and technology*, Lawrence Erlbaum, Mahwah, pp. 425–439, viewed 24 June 2017, <http://www.aect.org/edtech/edition3/ER5849x_C034.fm.pdf>.

52. DiBello, L, Lehman, D & Missildine, W 2010, How do you find an expert? identifying blind spots and complex mental models among key organizational decision makers using a unique profiling tool, in K Mosier and U Fischer (eds.), *Informed by Knowledge: Expert Performance in Complex Situations*, Psychology Press, New York, pp. 261–274, viewed 24 June 2017, <http://wtri.com/wp-content/uploads/2015/06/Informed-By-Knowledge-Chapter-12.pdf>.

53. DiBello, L & Missildine, W 2008, "Information technologies and intuitive expertise: a method for implementing complex organizational change among new yorkcity transit authority's bus maintainers," *Cognition, Technology and Work*, vol. 12, no. 1, pp. 61–75, http://dx.doi.org/10.1007/s10111-008-0126-z.

54. _____ 2011, "Future of immersive instructional design for the global knowledge economy: a case study of an ibm project management training in virtual worlds," *International Journal of Web Based Learning and Teaching Technologies*, vol. 6, no. 3, pp. 14–34, http://dx.doi.org/10.4018/jwltt,2011070102.

55. Dörfler, V, Baracskai, Z & Velencei, J 2009, "Knowledge levels: 3-d model of the levels of expertise," paper presented to the 68th Annual Meeting of the Academy of Management, Chicago, viewed 24 June 2017, <http://www.viktordorfler.com/webdav/papers/KnowledgeLevels.pdf>.

56. Drejer, A 2000, "Organisational learning and competence development," *The Learning Organization*, vol. 7, no. 4, pp. 206–220.

REFERENCES

57. Dreyfus, H 2005, "Overcoming the myth of the mental: how philosophers can profit from the phenomenology of everyday expertise," *Proceedings and Addresses of the American Philosophical Association*, November, Newark, DE, pp. 47-65. viewed <http://socs.berkeley.edu/~hdreyfus/pdf/Dreyfus%20APA%20Address%20%2010.22.05%20.pdf>.

58. Dreyfus, HL 2004, "A phenomenology of skill acquisition as the basis for a merleau-pontian non-representationalist cognitive science,", viewed 24 June 2017, <http://socrates.berkeley.edu/~hdreyfus/pdf/MerleauPontySkillCogSci.pdf>.

59. Dreyfus, HL & Dreyfus, SE 1980, *A Five-Stage Model of the Mental Activities Involved in Directed Skill Acquisition*, Report No. ORC 90-2, Operations Research Center, University of California, Berkeley, viewed 24 June 2017, <http://www.dtic.mil/cgi-bin/GetTRDoc?AD=ADA084551>.

60. _____ 1986, *Mind over machine: the power of human intuition and expertise in the era of the computer*, The Free Press, New York, http://dx.doi.org/10.1109/mex.1987.4307079.

61. _____ 2004, "The ethical implications of the five-stage skill-acquisition model," *Bulletin of Science, Technology and Society*, vol. 24, no. 3, pp. 251–264, http://dx.doi.org/10.1177/0270467604265023.

62. _____ 2005, "Peripheral vision: expertise in real world contexts," *Organization Studies*, vol. 26, no. 5, pp. 779–792, http://dx.doi.org/10.1177/0170840605053102.

63. _____ 2008, Beyond expertise: some preliminary thoughts on mastery, in K Nielsen (ed.), *A qualitative stance: Essays in honor of Stiener Kvale*, Aarhus University Press, Aarhus, Denmark, pp. 113–124, viewed 24 June 2017, <http://128.32.192.116/People/Faculty/dreyfus-pubs/mastery.doc>.

64. _____ 2009, The relationship of theory and practice in the acquisition of skill, in P Benner, C Tanner & C Chesla (eds.), *Expertise in nursing practice: Caring, clinical judgment, and ethics*, Springer, New York, pp. 1–23, viewed 24 June 2017, <http://lghttp.48653.nexcesscdn.net/80223CF/springer-static/media/samplechapters/9780826125446/9780826125446_chapter.pdf>.

65. Dror, IE 2011, The paradox of human expertise: why experts get it wrong, in N Kapur (ed.), *The paradoxical brain*, Cambridge University Press, New York, pp. 177–188, http://dx.doi.org/10.1017/cbo9780511978098.011.

66. Druckman, D & Bjork, RA 1991, *In the mind's eye: enhancing human performance*, D Druckman & R Bjork (eds.), National Academies Press, Washington, D.C., http://dx.doi.org/10.17226/1580.

67. Dunphy, BC & Williamson, SL 2004, "In pursuit of expertise: toward an educational model of expertise development," *Advances in Health Sciences Education*, vol. 9, no. 2, pp. 107–127, http://dx.doi.org/10.1023/B:AHSE.0000027436.17220.9c.

68. Eraut, M 1994, *Developing professional knowledge and competence*, Routledge, London, http://dx.doi.org/10.4324/9780203486016.

69. Ericsson, KA 2004, "Deliberate practice and the acquisition and maintenance of expert performance in medicine and related domains," *Academic Medicine*, vol. 79, no. 10, pp. 70–81, http://dx.doi.org/10.1097/00001888-200410001-00022.

70. _____ 2006, The influence of experience and deliberate practice on the development of superior expert performance, in K Ericsson, N Charness, P Feltovich & R Hoffman (eds.), *The Cambridge Handbook of Expertise and Expert Performance*, Cambridge University Press, New York, pp. 683–704, http://dx.doi.org/10.1017/CBO9780511816796.038.

71. _____ 2009a, "Discovering deliberate practice activities that overcome plateaus and limits on improvement of performance," in A Willamon, S Pretty & R Buck (eds.), *International Symposium on Performance Science*, Auckland, 15–18 December, European Association of Conservatoires (AEC), Utrecht, The Netherlands, pp. 11–21, viewed 24 June 2017, <http://www.performancescience.org/ISPS2009/Proceedings/Rows/003Ericsson.pdf>.

72. _____ 2009b, Enhancing the development of professional performance: implications from the study of deliberate practice, in K Ericsson (ed.), *Development of professional expertise: Toward measurement of expert performance and design of optimal learning environments*, Cambridge University Press, New York, pp. 405–431, http://dx.doi.org/10.1017/cbo9780511609817.022.

73. Ericsson, KA & Charness, N 1994, "Expert performance: its structure and acquisition," *American Psychologist*, vol. 49, no. 8, pp. 725–747, http://dx.doi.org/10.1037/0003-066X.49.8.725.

74. Ericsson, KA, Krampe, RTR, Tesch-romer, C, Ashworth, C, Carey, G, Grassia, J, Hastie, R, Heizmann, S, Kellogg, R, Levin, R, Lewis, C, Oliver, W, Poison, P, Rehder, R, Schlesinger, K, Schneider, V & Tesch-Römer, C 1993, "The role of deliberate practice in the acquisition of expert performance," *Psychological Review*, vol. 100, no. 3, pp. 363–406, http://dx.doi.org/10.1037/0033-295X.100.3.363.

75. Ericsson, KA & Lehmann, AC 1996, "Expert and exceptional performance: evidence of maximal adaptation to task constraints," *Annual Review of Psychology*, vol. 47, pp. 273–305, http://dx.doi.org/10.1146/annurev.psych.47.1.273.

76. Fadde, PJ 2009, "Training of expertise and expert performance," *Technology, Instructional, Cognition and Learning*, vol. 7, no. 2, pp. 77–81, viewed 24 June 2017, <http://web.coehs.siu.edu/Units/CI/Faculty/PFadde/Research/xbtintro.pdf>.

77. _____ 2010, "Look'ma, no hands: part-task training of perceptual-cognitive skills to accelerate psychomotor expertise," *The Interservice Industry Training, Simulation and Education Conference (I/ITSEC)*, Orlando, National Training and Simulation Association (NTSA), Arlington, pp. 1–10, viewed 24

June 2017, <http://ntsa.metapress.com/index/A4018183135VJ842.pdf>.

78. Fadde, PJ & Klein, G 2010, "Deliberate performance: accelerating expertise in natural settings," *Performance Improvement*, vol. 49, no. 9, pp. 5–14, http://dx.doi.org/10.1002/pfi.

79. Farr, MJ 1986, *The Long-Term Retention of Knowledge and Skills: A Cognitive and Instructional Perspective*, IDA Meorandum No. M-205, Insitute for Defence Analyses, Alexandria, viewed 24 June 2017, <http://www.dtic.mil/dtic/tr/fulltext/u2/a175905.pdf>.

80. Farrington-Darby, T & Wilson, JR 2006, "The nature of expertise: a review," *Applied Ergonomics*, vol. 37, no. 1, pp. 17–32, http://dx.doi.org/10.1016/j.apergo.2005.09.001.

81. Farrow, D & Abernethy, B 2002, "Can anticipatory skills be learned through implicit video-based perceptual training?," *Journal of Sports Sciences*, vol. 20, no. 6, pp. 471–85, http://dx.doi.org/10.1080/02640410252925143.

82. Feldon, DF, Timmerman, BC, Stowe, KA & Showman, R 2010, "Translating expertise into effective instruction: the impacts of cognitive task analysis (cta) on lab report quality and student retention in the biological sciences," *Journal of Research in Science Teaching*, vol. 47, no. 10, pp. 1165–1185, http://dx.doi.org/10.1002/tea.20382.

83. Feltovich, PJ, Spiro, RJ & Coulson, RL 1997, Issues of expert flexibility in contexts characterized by complexity and change, in P Feltovich, K Ford & R Hoffman (eds.), *Expertise in context: Human and machine*, MIT Press, Cambridge, pp. 125–146, viewed 24 June 2017, <http://www.academia.edu/download/29334639/Issues_of_Expert_Flexibili ty.pdf>.

84. Fitts, P & Posner, M 1967, *Learning and skilled performance in human performance*, Brock-Cole, Belmont.

85. Fitts, PM 1964, Perceptual-motor skill learning, in A Melton (ed.), *Categories of human learning*, Academic Press, New York, pp. 243–285, viewed 24 June 2017, <http://www.sciencedirect.com/science/article/pii/B9781483231457500169>.

86. Flyvbjerg, B 1990, "Rationalitet, intuition og krop i menneskets læreproces: fortolkning og evaluering af hubert og stuart dreyfus' model for indlæring af færdigheder (sustaining non-rationalized practices: body-mind, power, and situational ethics: an interview with hubert and stuart dreyfus),".

87. _____ 1991, "Sustaining non-rationalized practices: body-mind, power, and situational ethics. an interview with hubert and stuart dreyfus," *Praxis International*, no. 1, pp. 93–113.

88. Ge, X & Hardré, PL 2010, "Self-processes and learning environment as influences in the development of expertise in instructional design," *Learning Environments Research*, vol. 13, no. 1, pp. 23–41, http://dx.doi.org/10.1007/s10984-009-9064-9.

89. Glaser, R & Chi, MTH 1988, Overview, in Chi, Michelene TH and Glaser, Robert and Farr, MJ (eds.), *The nature of expertise*, Lawrence Erlbaum, Mahwah, pp. xv–xxviii.

90. Gobet, F 2013, "Expertise vs. talent," *Talent Development & Excellence*, vol. 5, no. 1, pp. 75–86, viewed <http://citeseerx.ist.psu.edu/viewdoc/download?doi=10.1.1.297.2267&rep=rep 1&type=pdf>.

91. Van Gog, T, Ericsson, KA, Rikers, RMJP & Paas, F 2005, "Instructional design for advanced learners: establishing connections between the theoretical frameworks of cognitive load and deliberate practice," *Educational Technology Research and Development*, vol. 53, no. 3, pp. 73–81, http://dx.doi.org/10.1007/BF02504799.

92. Greene, L, Lemieux, K & McGregor, R 1993, "Novice to expert: an application of the dreyfus model to management development in health care.," *Journal of Health and Human Resources Administration*, vol. 16, no. 1, p. 85.

93. Grenier, RS & Kehrhahn, M 2008, "Toward an integrated model of expertise redevelopment and its implications for hrd," *Human Resource Development Review*, vol. 7, no. 2, pp. 198–217, http://dx.doi.org/10.1177/1534484308316653.

94. De Groot, A 1965, *Thought and choice in chess (translated from the dutch original, 1946)*, Reprinted edn, G Baylor (ed.), Ishi Press, New York.

95. De Groot, A 1966, Perception and memory versus thought: some old ideas and recent findings, in B Kleinmuntz (ed.), *Problem solving*, John Wiley, New York, pp. 19–50.

96. Gunderman, RB 2009, "Competency-based training: conformity and the pursuit of educational excellence," *Radiology*, vol. 252, no. 2, pp. 324–326, http://dx.doi.org/10.1148/radiol.2522082183.

97. Guskey, T 2009, "Mastery learning,", viewed 24 June 2017, <http://www.education.com/reference/article/mastery-learning/>.

98. Guskey, TR & Gates, SL 1986, "Synthesis of research on the effects of mastery learning in elementary and secondary classrooms," *Educational Leadership*, pp. 73–80.

99. Hambrick, DZ, Altmann, EM, Oswald, FL, Meinz, EJ, Gobet, F & Campitelli, G 2014, "Accounting for expert performance: the devil is in the details," *Intelligence*, vol. 45, no. 4, pp. 112–114, http://dx.doi.org/10.1016/j.intell.2014.01.007.

100. Hambrick, DZ, Oswald, FL, Altmann, EM, Meinz, EJ, Gobet, F & Campitelli, G 2014, "Deliberate practice: is that all it takes to become an expert?," *Intelligence*, vol. 45, no. 4, pp. 34–45, http://dx.doi.org/10.1016/j.intell.2013.04.001.

101. Hesketh, B & Neal, A 1999, Technology and performance, in D Ilgen & E Pulakos (eds.), *The changing nature of performance: Implications for staffing,*

motivation, and development, Jossey-Bass, San Francisco, pp. 21–55.

102. Hoffman, R, Feltovich, PJ, Fiore, S, Klein, G & Moon, B 2008, *Program on Technology Innovation: Accelerating the Achievement of Mission-Critical Expertise: A Research Roadmap,* Report No. 1016710, Electric Power Research Institute (EPRI), Palo Alto, viewed 24 June 2017, <http://perigeantechnologies.com/publications/AcceleratingAchievementofExpertise.pdf>.

103. Hoffman, RR 1998, How can expertise be defined? implications of research from cognitive psychology, in R Williams, W Faulkner & J Fleck (eds.), *Exploring expertise,* Palgrave Macmillan, Edinburgh, Scotland, pp. 81–100, http://dx.doi.org/10.1007/978-1-349-13693-3_4.

104. Hoffman, RR (ed) 2007, *Expertise out of context: Proceedings of the sixth international conference on naturalistic decision making,* Taylor and Fracis, Boca Raton.

105. Hoffman, RR & Lintern, G 2006, Eliciting and representing the knowledge of experts, in K Ericsson, N Charness, P Feltovich & R Hoffman (eds.), *Cambridge handbook of expertise and expert performance,* Cambridge University Press, New York, pp. 203–222, http://dx.doi.org/10.1017/CBO9780511816796.012.

106. Hoffman, RR, Ward, P, Feltovich, PJ, DiBello, L, Fiore, SM & Andrews, DH 2014, *Accelerated Expertise: Training for high proficiency in a complex world,* Expertise: Research and Applications Series, Psychology Press, New York, http://dx.doi.org/10.4324/9780203797327.

107. Hunter, JE 1986, "Cognitive ability, cognitive aptitudes, job knowledge, and job performance," *Journal of Vocational Behavior,* vol. 29, no. 3, pp. 340–362, http://dx.doi.org/10.1016/0001-8791(86)90013-8.

108. Jacobs, R 2001, Managing employee competence and human intelligence in global organizations, in F Richter (ed.), *Maximizing human intelligence in Asia business: The sixth generation project. New York: Prentice-Hall,* Prentice-Hall, New York, pp. 44–54.

109. Jacobs, R & Hawley, J 2002, Emergence of workforce development: definition, conceptual boundaries, and future perspectives, in R MacLean & D Wilson (eds.), *International Handbook of Technical and Vocational Education and Training,* Kluwer, Amsterdam, viewed 24 June 2018, <http://www.economicmodeling.com/wp-content/uploads/2007/11/jacobs_hawley-emergenceofworkforcedevelopment.pdf>.

110. Jacobs, R & Washington, C 2003, "Employee development and organizational performance: a review of literature and directions for future research," *Human Resource Development International,* vol. 6, no. 3, pp. 343–354, http://dx.doi.org/10.1080/13678860110096211.

111. Jacobs, RL 1997, "A taxonomy of employee development: toward an organizational culture of expertise," *Proceedings of the Academy of Human Resource Development,* Academy of Human Resource Development, Baton

Rouge, pp. 278-283.

112. _____ 2003, *Structured on-the-job training: Unleashing employee expertise in the workplace*, 2nd edn, Berrett-Koehler, San Francisco.

113. Jacobson, MJ & Spiro, RJ 1991, "Hypertext learning environments and cognitive flexibility: characteristics promoting the transfer of complex knowledge," in L Birnbaum (ed.), *Proceedings of the International Conference on the Learning Sciences*, Association for the Advancement of Computing in Education, Charlottesville, pp. 240–245.

114. Jonassen, D, Ambruso, D & Olesen, J 1992, "Designing hypertext on transfusion medicine using cognitive flexibility theory," *Journal of Educational Multimedia and Hypermedia*, vol. 1, no. 3, pp. 309–322.

115. Jonassen, DH, Dyer, D, Peters, K, Robinson, T, Harvey, D, King, M & Loughner, P 1997, Cognitive flexibility hypertexts on the web: engaging learners in meaning making, in B Khan (ed.), *Web-based instruction*, Education Technology, Englewood Cliffs, pp. 119–133, viewed 24 June 2017, <http://www.academia.edu/download/6676532/36489.pdf>.

116. Kanfer, R & Kantrowitz, TM 2002, Ability and non-ability predictors of job performance, in S Sonnentag (ed.), *Psychological management of individual performance*, John Wiley, pp. 27–50, http://dx.doi.org/10.1002/0470013419.ch2.

117. Katz, SN, Hall, E & Lesgold, A 1997, *Cognitive Task Analysis and Intelligent Computer-Based Training Systems: Lessons Learned from Coached Practice Environments in Air Force Avionics*, Report No. TM 027360, University of Pittsburgh, Pittsburg, PA, viewed 24 June 2017, <https://archive.org/details/ERIC_ED411309>.

118. Khan, K & Ramachandran, S 2012, "Conceptual framework for performance assessment: competency, competence and performance in the context of assessments in healthcare-deciphering the terminology," *Medical Teacher*, vol. 34, no. 11, pp. 920–928, http://dx.doi.org/10.3109/0142159X.2012.722707.

119. Kim, JW, Ritter, FE & Koubek, RJ 2013, "An integrated theory for improved skill acquisition and retention in the three stages of learning," *Theoretical Issues in Ergonomics Science*, vol. 14, no. 1, pp. 22–37, http://dx.doi.org/10.1080/1464536X.2011.573008.

120. Kim, MK 2012, "Theoretically grounded guidelines for assessing learning progress: cognitive changes in ill-structured complex problem-solving contexts," *Educational Technology Research and Development*, vol. 60, no. 4, pp. 601–622, http://dx.doi.org/10.1007/s11423-012-9247-4.

121. _____ 2015, "Models of learning progress in solving complex problems: expertise development in teaching and learning," *Contemporary Educational Psychology*, vol. 42, no. 3, pp. 1–16, http://dx.doi.org/10.1016/j.cedpsych.2015.03.005.

122. Klein, G 1997, "Developing expertise in decision making," *Thinking \& Reasoning*, vol. 3, no. 4, pp. 337–352,

REFERENCES

http://dx.doi.org/10.1080/135467897394329.

123. Klein, GA 1998, *Sources of power: how people make decisions*, MIT Press, Cambridge.

124. Klein, GA & Baxter, HC 2009, Cognitive transformation theory: contrasting cognitive and behavioral learning, in D Schmorrow, J Cohn & D Nicholson (eds.), *The PSI handbook of virtual environment for training and education: Developments for the military and beyond, Volume 1, Education: Learning, requirements and metrics*, Praeger Security International, Santa Barbara, pp. 50–65, viewed 24 June 2017, <https://pdfs.semanticscholar.org/99f0/b9bdbce6432c3232fdeffeae0fddea7b cebd.pdf>.

125. Klein, GA & Hoffman, RR 1993, "Perceptual-cognitive aspects of expertise" M Robinowixz (ed.), *Cognitive Science Foundations of Instruction*, pp. 203–226, viewed 24 October 2018, <http://cmapsinternal.ihmc.us/rid=1217527241618_235135118_3994/Seeing %20the%20Invisible-1992.pdf>.

126. Kuchenbrod, R 2016, "Accelerating expertise to facilitate decision making in high-risk professions using the DOCUM system," PhD thesis, Eastern Illinois University, Charleston, viewed 24 June 2017, <http://thekeep.eiu.edu/cgi/viewcontent.cgi?article=3462&context=theses>.

127. Kulasegaram, KM, Grierson, LEM & Norman, GR 2013, "The roles of deliberate practice and innate ability in developing expertise: evidence and implications," *Medical Education*, vol. 47, no. 10, pp. 979–989, http://dx.doi.org/10.1111/medu.12260.

128. Lajoie, SP 2003, "Transitions and trajectories for studies of expertise," *Educational Researcher*, vol. 32, no. 8, pp. 21–25, http://dx.doi.org/10.3102/0013189X032008021.

129. Langan-Fox, J, Armstrong, K, Balvin, N & Anglim, J 2002, "Process in skill acquisition: motivation, interruptions, memory, affective states, and metacognition," *Australian Psychologist*, vol. 37, no. 2, pp. 104–117, http://dx.doi.org/10.1080/00050060210001706746.

130. Lombardo, MP & Deaner, RO 2014, "You can't teach speed: sprinters falsify the deliberate practice model of expertise," *PeerJ*, vol. 2: e445, pp. 1–31, http://dx.doi.org/10.7717/peerj.445.

131. London, M & Mone, EM 1999, Continuous learning, in D Ilgen & E Pulakos (eds.), *The changing nature of performance: Implications for staffing, motivation, and development*, Jossey-Bass, San Francisco, pp. 119–153.

132. Macmillan, PJ 2015, "Thinking like an expert lawyer: measuring specialist legal expertise through think-aloud problem solving and verbal protocol analysis," PhD thesis, BOND UNIVERSITY, Robina, Australia, viewed 24 June 2017, <http://epublications.bond.edu.au/cgi/viewcontent.cgi?article=1167&context= theses>.

133. McClelland, DC, Atkinson, JW, Clark, RA & Lowell, EL 1976, *The achievement motive.*, Irvington, New York.

134. McDaniel, MA, Schmidt, FL & Hunter, JE 1988, "Job experience correlates of job performance," *Journal of Applied Psychology*, vol. 73, no. 2, p. 327, http://dx.doi.org/10.1037/0021-9010.73.2.327.

135. McElroy, E, Greiner, D & de Chesnay, M 1991, "Application of the skill acquisition model to the teaching of psychotherapy," *Archives of Psychiatric Nursing*, vol. 5, no. 2, pp. 113–117.

136. Merkelbach, EJHM & Schraagen, JMC 1994, *A Framework for the Analysis of Cognitive Tasks*, Report No. TNO-TM 1994 B-13, TNO Human Factors Research, Soesterberg, The Netherlands, viewed 24 June 2017, <http://www.dtic.mil/docs/citations/ADA285345>.

137. Van Merriënboer, JJ, Jelsma, O & Paas, FG 1992, "Training for reflective expertise: a four-component instructional design model for complex cognitive skills," *Educational Technology Research and Development*, vol. 40, no. 2, pp. 23–43, http://dx.doi.org/10.1007/BF02297047.

138. Van Merriënboer, JJ & Kester, L 2008, Whole-task models in education, in J Spector, M Merrill, J van Merriënboer & M Driscoll (eds.), *Handbook of research on educational communications and technology*, Erlbaum/Routledge, Mahwah, pp. 441–456, viewed 24 June 2017, <https://www.researchgate.net/publication/268000667>.

139. Van Merriënboer, JJG, Clark, RE & de Croock, MBM 2002, "Blueprints for complex learning: the 4c/id-model," *Educational Technology Research and Development*, vol. 50, no. 2, pp. 39–61, http://dx.doi.org/10.1007/BF02504993.

140. Merrill, MD 2002, "First principles of instruction," *Educational Technology Research and Development*, vol. 50, no. 3, pp. 43–59, http://dx.doi.org/10.1007/bf02505024.

141. _____ 2006, Hypothesized performance on complex tasks as a function of scaled instructional strategies, in J Enen & R Clark (eds.), *Handling complexity in learning environments: Research and theory*, Elsevier, Amsterdam, pp. 265–282, viewed 24 June 2017, <http://www.mdavidmerrill.com/Papers/Scaled_Instructional_Strategies.pdf>.

142. Moon, YK, Kim, EJ & You, Y-M 2013, "Study on expertise development process based on arête," *International Journal of Information and Education Technology*, vol. 3, no. 2, pp. 226–230, http://dx.doi.org/10.7763/IJIET.2013.V3.269.

143. Motowidlo, SJ, Borman, WC & Schmit, MJ 1997, "A theory of individual differences in task and contextual performance," *Human Performance*, vol. 10, no. 2, pp. 71–83, http://dx.doi.org/10.1207/s15327043hup1002_1.

144. Novak, D 2011, *The systematic development of expertise*, CreateSpace Independent Publishing Platform, USA.

REFERENCES

145. Oyewole, S a, Farde, AM, Haight, JM & Okareh, OT 2011, "Evaluation of complex and dynamic safety tasks in human learning using the act-r and soar skill acquisition theories," *Computers in Human Behavior*, vol. 27, no. 5, pp. 1984–1995, http://dx.doi.org/10.1016/j.chb.2011.05.005.

146. Page-Jones, M 1998, "The seven stages of expertise in software engineering,", viewed 24 October 2018, <http://www. waysys. com/ws_content_al_sse. html>.

147. Peña, A 2010, "The dreyfus model of clinical problem-solving skills acquisition: a critical perspective," *Medical Education Online*, vol. 15, no. 1, pp. 1–11, http://dx.doi.org/10.3402/meo.v15i0.4846.

148. Qui'nones, MA, Ford, JK & Teachout, MS 1995, "The relationship between work experience and job performance: a conceptual and meta-analytic review," *Personnel Psychology*, vol. 48, no. 4, pp. 887–910, http://dx.doi.org/10.1111/j.1744-6570.1995.tb01785.x.

149. Ramsburg, L 2010, "An initial investigation of the applicability of the dreyfus skill acquisition model to the professional development of nurse educators," PhD thesis, Marshall University Graduate College, Huntington, viewed 24 June 2017, <http://mds.marshall.edu/cgi/viewcontent.cgi?article=1371&context=etd>.

150. Rasmussen, J 1983, "Skills, rules, and knowledge; signals, signs, and symbols, and other distinctions in human performance models," *IEEE Transactions on Systems, Man, and Cybernetics*, no. 3, pp. 257–266, viewed <http://www.newviewofhealthandsafety.com/s/SkillsRulesAndKnowledge-Rasmussen.pdf>.

151. Rasmussen, J, Pejtersen, AM & Goodstein, L 1994, *Cognitive systems engineering*, John Wiley, New York.

152. Reigeluth, C 1999, "What is instructional design theory and how is it changing?" C Reigeluth (ed.), *Instructional-Design Theories and Models: A New Paradigm of Instructional Theory*, Vol. 2, pp. 5–29.

153. Robinson, WL 1974, "Conscious competency - mark of a competent instructor," *The Personnel Journal*, pp. 538–539.

154. Rosenberg, M 2012, "Beyond competency: it's the journey to mastery that counts," *Learning Solutions Magazine*, viewed 24 June 2017, <http://www.learningsolutionsmag.com/articles/930/beyond-competence-its-the-journey-to-mastery-that-counts>.

155. Roth, Milie M & O'Hara, J 2014, "Discussion panel: how to recognize a "good" cognitive task analysis?," *Proceedings of the Human Factors and Ergonomics Society 58th Annual Meeting*, Chicago, 27-31 October, SAGE, Thousand Oaks, pp. 320–324, http://dx.doi.org/10.1177/1541931214581066.

156. Salas, E, Cannon-Bowers, J, Church-Payne, S & Smith-Jentsch, K 1998, "Teams and teamwork in the military" C Cronin (ed.), *Military Psychology: An Introduction*, pp. 71–87.

157. Schaafstal, AA, Schraagen, JM, van Berlo, M & van Berlo, M 2000, "Cognitive task analysis and innovation of training: the case of structured troubleshooting," *Human Factors*, vol. 42, no. 1, pp. 75–86, http://dx.doi.org/10.1518/001872000779656570.

158. Schmidt, FL, Hunter, JE & Outerbridge, AN 1986, "Impact of job experience and ability on job knowledge, work sample performance, and supervisory ratings of job performance.," *Journal of Applied Psychology*, vol. 71, no. 3, pp. 432–439, http://dx.doi.org/0021-9010.71.3.432.

159. Schmitt, J 1996, "Designing good TDGS," *Marine Corps Gazette*, vol. 80, no. 5, pp. 96–97.

160. Schneider, W 1993, Acquiring expertise: determinants of exceptional performance., in K Heller, F Mönks & A Passow (eds.), *International handbook of research and development of giftedness and talent*, Pergamon Press, Elmsford, NY, pp. 311–324, viewed <https://opus.bibliothek.uni-wuerzburg.de/frontdoor/deliver/index/docId/7140/file/Schneider_W_OPUS_7140.pdf>.

161. Schreiber, BT, Bennett Jr, W, Colegrove, CM, Portrey, AM, Greschke, DA & Bell, HH 2009, Evaluating pilot performance, in K Ericsson (ed.), *Development of professional expertise*, Cambridge University Press, New York, pp. 247–270, http://dx.doi.org/10.1017/cbo9780511609817.014.

162. Scobey, BW 2006, "The journey to expertise: pathways to expert knowledge traveled by texas juvenile probation officers," PhD thesis, Texas State University, San Marcos, viewed 24 June 2017, <https://digital.library.txstate.edu/bitstream/handle/10877/4267/fulltext.pdf?sequence=1>.

163. Seamster, TL, Redding, RE, Cannon, JR, Ryder, JM & Purcell, JA 1993, "Cognitive task analysis of expertise in air traffic control," *The International Journal of Aviation Psychology*, vol. 3, no. 4, pp. 257–283.

164. Shiffrin, RM & Schneider, W 1977, "Controlled and automatic human information processing: perceptual learning, automatic attending and a general theory," *Psychological Review*, vol. 84, no. 2, pp. 127–190, viewed 24 June 2017, <http://www.bryanburnham.net/wp-content/uploads/2014/01/Shiffrin-1977-Psychological-Review.pdf>.

165. Shuell, TJ 1986, "Individual differences: changing conceptions in research and practice," *American Journal of Education*, vol. 94, no. 3, pp. 356–377, http://dx.doi.org/10.1086/443854.

166. Simpson, PA 2001, *Naturalistic Decision Making in Aviation Environments*, Research Report No. DSTO-GD-0279, DSTO Aeronautical and Maritime Research Laboratory, Fisherman's Bend, Victoria, Australia, viewed 24 October 2018, <http://www.dtic.mil/get-tr-doc/pdf?AD=ADA393102>.

167. Soderstrom, NC & Bjork, RA 2015, "Learning versus performance: an integrative review," *Perspectives on Psychological Science*, vol. 10, no. 2, pp. 176–199, http://dx.doi.org/10.1177/1745691615569000.

168. Sonnentag, S & Frese, M 2002, Performance concepts and performance theory, in S Sonnentag (ed.), *Psychological management of individual performance*, John Wiley, New York, pp. 3–25, http://dx.doi.org/10.1002/0470013419.ch1.

169. Sonnentag, S & Kleine, BM 2000, "Deliberate practice at work: a study with insurance agents," *Journal of Occupational and Organizational Psychology*, vol. 73, no. 1, pp. 87–102, http://dx.doi.org/10.1348/096317900166895.

170. Soule, RT 2016, "The learning experience of tough cases: a descriptive case study," PhD thesis, The George Washington University, Ann Harbor, MI, viewed 24 June 2017, <http://pqdtopen.proquest.com/doc/1751007250.html?FMT=AI>.

171. Spiro, RJ, Collins, BP, Thota, JJ & Feltovich, PJ 2003, "Cognitive flexibility theory: hypermedia for complex learning, adaptive knowledge application, and experience acceleration," *Educational Technology*, vol. 43, no. 5, pp. 5–10.

172. Spiro, RJ, Coulson, RL, Feitovich, PJ & Anderson, DK 1988, *Cognitive Flexibility Theory: Advanced Knowledge Acquisition in Ill-Structured Domains*, Technical Report No. 441, University of Illinois, Champaign, viewed 24 June 2017, <http://files.eric.ed.gov/fulltext/ED302821.pdf>.

173. Spiro, RJ, Feltovich, PJ, Coulson, RL, Jacobson, M, Durgunoglu, A, Ravlin, S & Jehng, J-C 1992, *Knowledge Acquisition for Application: Cognitive Flexibility and Transfer of Training in Iii-Structured Domains*, ARI Research Note No. 92-21, United States Army Research Institute for the Behaviorai and Social Sciences, Alexandria, viewed 24 June 2017, <http://www.dtic.mil/dtic/tr/fulltext/u2/a250147.pdf>.

174. Spiro, RJ & Jehng, J-C 1990, Cognitive flexibility and hypertext: theory and technology for the nonlinear and multidimensional traversal of complex subject matter, in D Nix & R Spiro (eds.), *Cognition, education, and multimedia: Exploring ideas in high technology*, Erlbaum, Hilldale, pp. 163–205.

175. Spiro, RJ, Vispoel, WP, J G Schmitz, Samarapungavan, A & Boerger, AE 1987, *Knowledge Acquisition for Application: Cognitive Flexibility and Transfer in Complex Content Domains*, Report No. 409, University of Illinois, Champaign, viewed 24 June 2017, <http://files.eric.ed.gov/fulltext/ED287155.pdf>.

176. Squires, A, Wade, J, Dominick, P & Gelosh, D 2011, *Building a Competency Taxonomy to Guide Experience Acceleration of Lead Program Systems Engineers*, ERIC No. ADA589178, Stevens Institute of Technology, Hoboken, viewed 24 June 2017, <http://www.dtic.mil/cgi-bin/GetTRDoc?AD=ADA589178>.

177. Stefanidis, D, Korndorffer, JR, Markley, S, Sierra, R & Scott, DJ 2006, "Proficiency maintenance: impact of ongoing simulator training on laparoscopic skill retention," *Journal of the American College of Surgeons*, vol. 202, no. 4, pp. 599–603, http://dx.doi.org/10.1016/j.jamcollsurg.2005.12.018.

178. Sternberg, R & Spear-Swerling, L 1996, "Teaching for thinking. washington, dc, us: american psychological association,".

179. Sternberg, RJ 1988, *The triarchic mind: A new theory of human intelligence*, Viking-Penguin, New York.

180. _____ 1998a, "Metacognition, abilities, and developing expertise: what makes an expert student?," *Instructional Science*, vol. 26, no. 1-2, pp. 127–140, http://dx.doi.org/10.1023/A:1003096215103.

181. _____ 1998b, "Abilities are forms of developing expertise," *Educational Researcher*, vol. 27, no. 3, pp. 11–20.

182. _____ 1999, "Intelligence as developing expertise.," *Contemporary Educational Psychology*, vol. 24, no. 4, pp. 359–375, http://dx.doi.org/10.1006/ceps.1998.0998.

183. Sternberg, RJ & others 1985, *Beyond IQ: A triarchic theory of human intelligence*, Cambridge University Press, New York.

184. Sternberg, RJ & Zhang, L 2005, "Styles of thinking as a basis of differentiated instruction," *Theory into Practice*, vol. 44, no. 3, pp. 245–253, viewed 24 October 2018, <https://hub.hku.hk/bitstream/10722/45445/1/128798.pdf?accept=1>.

185. Stewart, J & Dohme, JA 2005, "Automated hover trainer: simulator-based intelligent flight training system," *International Journal of Applied Aviation Studies*, vol. 5, no. 1, pp. 25–40, viewed 24 October 2018, <https://pdfs.semanticscholar.org/b959/00872b74560e59c33864e1a7642aaf bbe23c.pdf#page=25>.

186. VanLehn, K 1996, "Cognitive skill acquisition," *Annual Review of Psychology*, vol. 47, no. 1, pp. 513–539, viewed 24 June 2017, <http://citeseerx.ist.psu.edu/viewdoc/download?doi=10.1.1.74.7707&rep=rep1 &type=pdf>.

187. VanSledright, B & Alexander, P 2002, "Historical knowledge, thinking, and beliefs: evaluation component of the corps of historical discovery project (# s215x010242),".

188. Velmahos, GC, Toutouzas, KG, Sillin, LF, Chan, L, Clark, RE, Theodorou, D & Maupin, F 2004, "Cognitive task analysis for teaching technical skills in an inanimate surgical skills laboratory," *The American Journal of Surgery*, vol. 187, no. 1, pp. 114–119, http://dx.doi.org/10.1016/j.amjsurg.2002.12.005.

189. Vygotsky, LS 1978, *Mind in society: the development of higher mental process*, Harvard University Press, Cambridge.

190. Ward, P, Farrow, D, Harris, KR, Williams, AM, Eccles, DW & Ericsson, KA 2008, "Training perceptual-cognitive skills: can sport psychology research inform military decision training?," *Military Psychology*, vol. 20, no. Suppl 1, pp. S71–S102, http://dx.doi.org/10.1080/08995600701804814.

191. Williams, AM, Ward, P & Chapman, C 2003, "Training perceptual skill in field hockey: is there transfer from the laboratory to the field?," *Research Quarterly for Exercise and Sport*, vol. 74, no. 1, pp. 98–103,

REFERENCES

http://dx.doi.org/10.1080/02701367.2003.10609068.

192. Woolley, NN & Jarvis, Y 2007, "Situated cognition and cognitive apprenticeship: a model for teaching and learning clinical skills in a technologically rich and authentic learning environment," *Nurse Education Today*, vol. 27, no. 1, pp. 73–9, http://dx.doi.org/10.1016/j.nedt.2006.02.010.

INDEX

INDEX

Speed To Proficiency
RESEARCH

Accelerated Performance for Accelerated Times

S2Pro© Speed To Proficiency Research is a corporate research and consulting forum that provides authentic guidelines to business practitioners to accelerate proficiency of their workforce, teams, and professionals at the 'speed of business'. S2Pro© publishes reports, ebooks, and articles exclusively related to accelerated performance, accelerated proficiency and accelerated expertise in the individual and organizational context. Our extensive knowledge base of "how to methods" is derived from experience-based and practice-based observations, analysis/synthesis of existing research, or based on planned/focused research studies through a network of researchers who exclusively focus on 'time' and 'speed' metrics in the business context.

Speed To Proficiency Research: S2Pro©, Singapore

Website: https://www.speedtoproficiency.com
e-mail: contact@speedtoproficiency.com
Facebook: https://www.facebook.com/speedtoproficiency/
LinkedIn: https://www.linkedin.com/company/speedtoproficiency/
Twitter: https://www.twitter.com/speed2expertise

www.ingramcontent.com/pod-product-compliance
Lightning Source LLC
Chambersburg PA
CBHW021432180326
41458CB00001B/233